The Human Body

Anatomical Terms and
What They Mean

2nd edition and revised

The Human Body

Anatomical Terms and What They Mean

2nd edition and revised

Professor Terry M. Mayhew

Emeritus Professor, School of Life Sciences,
Queen's Medical Centre, University of Nottingham

iooto9o35o

Copyright © Terry M. Mayhew, 2013

Second Edition 2013

Published by 5m Publishing
Benchmark House
8 Smithy Wood Drive
Sheffield
S35 1QN
United Kingdom

www.5mpublishing.com
books@5mpublishing.com

ISBN: 978-0-9555011-7-3

Printed and bound in the UK by Berforts Information Press.

Table of contents

Preface to First Edition

My family name is Mayhew and derived from the name Matthias which, in Hebrew, means 'gift of God'. This meaning may still hold appeal for my parents but I doubt whether any of my students see me in this light! The point is, although Mayhew has a meaning beyond merely identifying me as a member of this family line, that meaning has been lost over time. Occasionally, the meaning of a name has remained transparent. For example, we still recognise Smith as an occupational name even though its present owners may not pursue that occupation. In a sense, this book is about uncovering the mysteries of words and names.

An enterprise such as this may be justified in various ways. Basically, I enjoy words and their meanings and, for me, this has been justification enough. However, there is a more practical reason. Over the past 20 or so years in which I have taught Anatomy to medical and science students, it has become increasingly obvious that students are losing sight of the linguistic context in which anatomical terms and English words exist. In the case of students for whom English is the Mother tongue, the loss is explicable by the decline in teaching of Latin and Greek, languages whose influences on English and anatomical terminology are enormous. For overseas students in the UK (most of whom are from the Middle and Far East), the meaning and significance of anatomical terms may be even more opaque. Often, this is because they have less cultural exposure to Latin and Greek or even to the Romance languages (French, Italian, Spanish) which are, themselves, heavily influenced by Latin.

Here is a real example: I once examined a student from the Far East on the anatomy of the upper limb and, whilst she identified the structures which I pointed out, she was exceptionally slow in naming them. After a few minutes, I stopped the exam to enquire what the problem was. She responded that, having identified a structure, she next had to translate a mnemonic into the proper name. To understand her process, I asked her to talk me through how she identified (correctly) the **extensor carpi radialis longus** muscle. She recalled this name via the mnemonic **E**lephants **C**an **R**emember **L**atin and then had to decipher **E** for extensor, **C** for carpi and so on! In other words, **extensor carpi radialis longus** had as much explanatory significance to her as elephants can remember Latin! We resumed the viva, giving her a little more time to answer each spotter question, and she passed comfortably.

To those who, like me, have had a minimal or amateurish exposure to Latin or Greek, much of anatomical terminology has a layer of meaning which is transparent from the words themselves. Thus, the student above would have found it much easier and quicker to link the structure and name if, on observing two extensor muscles (**extensor**) of the wrist (**carpi**) on the same side of the forearm as the radius (**radialis**), she noted that one (**longus**) was longer than the other. In similar fashion, the name **latissimus dorsi** muscle should tell the reader that this is a very extensive (**latissimus**) muscle of the back (**dorsi**). The term **cricoid** should tell the reader that this laryngeal cartilage is ring (**cric-**) shaped (**-oid**). Actually, it has the shape of a signet-ring, with an anterior band and a posterior plate. The name **azygos** vein should inform that this vein on the right side has no (**a-**) pair or partner (**-zygos**) on the left side of the posterior thoracic wall.

Sometimes, anatomical terminology lets us down because of sloppy description, misspelling, abbreviation or the inevitable advances of time or technology. As an example of sloppy application, note that a cerebral **hemisphere** does <u>not</u> have the shape of a half-sphere; contrary to what its name suggests, the **innominate** bone of the pelvis - no (**in-**) name (**-nominate**) - <u>does</u> have a name! A widespread but disastrous misspelling is that of **fetus**: the incorrect **foetus** means 'something that stinks' but the correct **fetus** means 'something brought forth in pregnancy'. For an example of abbreviation, note that **skeleton** is short for **skeleton soma** which means 'the dry part of the body'. Finally, nowadays, it is not obvious why a **clavicle** was named because of its resemblance to a little key or why a **fibula** resembles a pin.

In an attempt to compensate for some of this obscurity, I have compiled a **glossary** (from Latin **glossa** 'word requiring explanation' and Greek **glossa** 'tongue', the same word-root which gives **hypoglossal** nerve, **glossopharyngeus** muscle, etc). The glossary is not exhaustive but embraces the most common and useful anatomical names, prefixes and suffixes. Emphasis is given to words which assist understanding. Therefore, words like artery (Latin for 'artery') and pharynx (Greek for 'pharynx') are excluded because they offer no added value. To examples of anatomical usage of a word-root, I have added other (often everyday) words of similar derivation just to emphasise that these "dead" languages are alive and well in modern English and in other areas of science. I very much hope that this compendium will allow students of Anatomy to make more sense of, and take more delight in, this fascinating and crucial activity. Note I said activity and not subject. **Anatomy** is something to be performed - it means cutting (**-tomy**) up (**ana-**)!

Preface to Second Edition

I am grateful to colleagues and reviewers for their comments on the first edition. Partly in response, I have added to Glossary 1 some new prefixes and suffixes which are of immediate relevance or may also help in understanding terms used in other areas including histology. In Glossary 2, as well as correcting some minor factual errors, several main terms have been expanded and new terms have been included. As in the first edition, I have deliberately avoided using eponyms because these are applied inconsistently between countries and, more importantly, they do not assist understanding.

Acknowledgements

As in the first edition, I dedicate this little book to all my students, past and present, who have encouraged me to learn more about the 'naming of parts' and 'meanings of words'. I am grateful to my Latin master at School, the late Mr JE Taylor, and I thank my mentors, the late Prof Robert Barer (Sheffield) and Prof John Clegg (Aberdeen), who fostered my learning of Anatomy in their Departments. Thanks also to my friends, Bob Banks (Durham), Reidar Myklebust (Bergen), Albrecht Reith (Oslo) and Mick Turton (Sheffield), who share my love of words and/or stimulated my interest in Latin, Greek and Norwegian. Finally, for their love and sufferance, I thank my wife (Joan), my family (Nick, Gemma, Zoë), my gorgeous grand-children (Emma, Charlie, Freddie, Shauna) and my late parents (Doris, Leslie).

Abbreviations

Arab: Arabic
Fr: French
Gael: Gaelic
Gr: Greek
It: Italian
L: Latin
ML: Medieval Latin
OE: Old English
ON: Old Norse or Norwegian
Sp: Spanish

a: artery; aa: arteries
CNS: central nervous system
IVC: inferior vena cava
lig: ligament; ligs: ligaments
m: muscle; mm: muscles
n: nerve; nn: nerves
SVC: superior vena cava
v: vein; vv: veins

References

I have found the following texts extremely useful:

Byrne M. (1987) *Eureka!*, Guild Publishing, London.

Holmen M. (ed.) (1999) *Engelsk Lommeordbok*, Kunnskapforlaget, Oslo.

Kidd D.A. (1984) *Collins Gem Latin Dictionary*, Collins, London and Glasgow.

MacFarlane P. (1955) *Pocket Dictionary of the Latin & English Languages*, Eyre and Spottiswoode, London.

Pearsall J. (ed.) (1998) *The New Oxford Dictionary of English*, Oxford University Press, Oxford.

Watts N. and George-Papageorgiou H. (1997) *Collins Pocket Greek Dictionary*, Harper-Collins, Glasgow.

Williams P.L., Warwick R., Dyson M. and Bannister L.H. (eds.) (1989) *Gray's Anatomy*, 37th edition, Churchill Livingstone, Edinburgh.

Zuckerman S. (1984) *A New System of Anatomy. A Dissector's Guide and Atlas*, Oxford University Press, Oxford.

Glossary 1

Some Prefixes and Suffixes Worth Remembering

A-, **An-**: [Gr] prefix negating what follows. For example, someone who is not political is said to be apolitical. As an anatomical example, the azygos v (draining intercostal spaces on the right side of the posterior thoracic wall) has no equivalent structure on the left (azygos means 'having no pair or partner'). Intercostal spaces on the left drain instead into a set of hemiazygos vv.

Ab-: [L] denotes *from, away from, the opposite*. Something 'opposite to normal' is abnormal; someone who abstains will 'refrain from' doing something; to absorb is to 'suck [liquid] from'. In Anatomy, moving the upper limb away from the side of the body is abduction (abduction means 'drawing away from').

Ad-: [L] signifies *towards, near, next to*. Adjacent means 'lying close by'; ad infinitum means 'going towards infinity' or 'going on forever' (like one of my lectures!). In anatomical terms, moving the upper limb towards the side of the body is adduction (adduction means 'drawing towards'); the adrenal glands really are 'near the kidneys'.

Ana-: [Gr] prefix indicating *up*. To perform an anatomy ('cutting up') originally meant to perform a dissection; anabolism ('building up') is constructive metabolism in which complex molecules are synthesised from simpler precursors, e.g. anabolic steroids promote the build-up of muscle.

Ante-: [L] signifies *before* (in time, space or comparison), *in front*. An antenatal visit occurs 'before birth'; ante meridiem (am) means 'before midday'. In Anatomy, anterior means 'in or at the front'; the antebrachium is the 'forearm'; anteflexion of the uterus is 'forward flexion'.

Anti-, **Ant-**: [Gr] denotes *against, opposing, opposite*. Anticlockwise is 'opposite of clockwise'; an antiperspirant 'opposes sweating'; an antibody acts 'against a body [an antigen]'. In Anatomy, an antagonist ('fighter against') is a muscle with an opposite effect or action, e.g. triceps brachii m (elbow extensor) is an antagonist of biceps brachii m (elbow flexor).

Apo-: [Gr] means *from, away, by, off*. An apostle is a 'person sent away [to preach the gospel]'; an apothecary 'puts away or stores' drugs. An aponeurosis (literally, 'from muscle' or 'from tissue') is a flattened tendon arising from a muscle, e.g. the bicipital aponeurosis from biceps brachii m; during apocrine ('breaking off') secretion, pieces of cytoplasm are lost, e.g. secretion of milk by mammary gland cells.

Arthro-, **Arthr-**: [Gr] means *related to joints*. Arthrology is 'the study of joints'. Arthritis is joint inflammation; arthropods are invertebrates having segmented bodies and jointed limbs.

Auto-: [Gr] meaning *self, own, same*. An automobile should be 'self-moving' (and not require a push or pull!); it may be an automatic 'acting of itself'; cell autolysis is 'self destruction'. The autonomic ('following its own laws') nervous system is involuntary, operating independently of the will.

Bi-, **Bis-**: [L] indicates *two, twice*. A bicycle has 'two wheels'; biscuit means 'twice-cooked' (originally this involved an initial bake followed by drying in a slow oven). The spinous process of a cervical vertebra is bifid ('split into two parts'). The biceps brachii m has 'two heads'; the bicuspid valve has 'two points' like a mitre, the hat of a bishop (hence the alternative name, mitral valve!). Related to **Di-** (see below).

Circum-: [L] means *round, about*. Circumcision involves 'cutting around'; the circumference 'passes around' a circle; circumnavigators 'sail around' the World. In Anatomy, circumduction at the shoulder joint is a movement in which the limb is 'drawn around' in a circle with successive flexion, abduction, extension and adduction at the joint; circumflex means 'bending around', e.g. the circumflex branch of the left coronary a bends around the left border of the heart to pass posteriorly.

-cle, -culus, -cula, -culum: [L] suffixes denoting *small, little*. A funicular ('using little ropes') railway runs up and down a slope using cables which counterbalance the carriages going up and down; cells have an endoplasmic reticulum ('little network') for protein synthesis and export. Similarly, in Anatomy, we have auricle ('little ear'), follicle ('small bag'), colliculus ('little hill'), retinacula ('little stays or ties' to bind down tendons) and trabecula ('little beam or strut').

-crine: [Gr] denotes *secretion, secreting*. Endocrine glands secrete into the blood; holocrine secretion involves loss of most of the secreting cell (as happens in secretion of sebum by sebaceous gland cells).

De-: [L] prefix indicating *from, away or reversal*. Defaecation ('getting rid of faeces') involves evacuating the bowels. A defibrillator is used in order to stop cardiac fibrillation

Di-: [Gr] prefix indicating *two, twice*. A dioxide has 'two oxygen atoms'; dichromatic means having 'two colours'. In Anatomy, the digastricus m possesses 'two bellies'. Related to **Bi-, Bis-** (see above).

Dia-: [Gr] means *across, through, between*. The diameter of a circle or sphere is the 'right across measurement'; it is no coincidence that we say of a particular food "It went straight through me!" – diarrhoea means, literally, 'a flowing through'. The diaphragm ('fence between') separates the thorax from the abdomen; the diaphysis of a long bone is its shaft – diaphysis means 'between the growth regions'. See **epi-** below.

Dis-: [L] means *apart, asunder*. The process of dissection involves the body being 'cut asunder'; disarticulation involves separating the bones that contribute to a joint.

Endo-: [Gr] indicates *within*. An endoskeleton is inside the body; an endocrine ('secreting inwards') gland secretes into the bloodstream; the endoderm ('inner skin or layer') is the innermost germ layer; endocardium ('innermost heart [region]') really is the innermost layer of the heart.

Epi-: [Gr] denotes *on, near, upon, above, over, after*. Ephemeral means 'lasting only one day'; an epidemic ('on the people') affects many people in a given area. The epiphysis ('on the growth region') of a long bone lies on the growth plate and underlying growth region (metaphysis); epithelium (literally 'on the nipple') actually covers many other surfaces as well!

Exo-, Ex-, Ecto-: [Gr] means *external, outside, beyond*. The Exodus involved the Israelites finding a 'way out' of Egypt; exotic means 'out of the ordinary'; exogenous means 'having an outside origin'. Expiration is the 'breathing out' phase of mechanical respiration. Ectoderm is the outermost of the primary germ layers; an ectopic pregnancy is one which does not occur in the uterus – so it is 'out of place'.

Hemi-: [Gr] signifies *half*. A hemisphere is 'half a sphere' but, unfortunately, this does not describe a cerebral hemisphere! See **semi-** below.

Hyper-: [Gr] means *above, over, in excess*. A hypercritical individual is 'overly critical'; a child who is "hyper" (short for hyperactive) is 'overactive'; a hypermarket is above an ordinary market (at least in size). Hyperplasia ('excessive creation') is cell division above the normal rate.

Hypo-: [Gr] signifies *below, beneath, down*. A hypocrite is 'not critical enough' (he/she doesn't include themselves!); formulating a hypothesis ('placing down') involves setting down an idea for testing. In Anatomy, the hypophysis (pituitary) arises as a 'downgrowth' from the hypothalamus ('below the thalamus'); in at least part of its course, the hypoglossal n is seen running 'below the tongue'.

-iform: [L] suffix indicating *of a certain shape or form*. Same as **–oid** [Gr] (see below). Falciform ('sickle-shaped') ligament; fusiform ('spindle-shaped'); piriform ('pear-shaped') recess; pisiform ('pea-shaped') bone.

Infra-, infero-: [L] prefix signifying *below, beneath, after*. Something infra dig is 'below one's dignity'; infrastructure is the 'underlying structure' of a system. The infraglenoid tubercle lies 'below the glenoid cavity'; the infraspinatus m lies 'below the spine [of the scapula]'; inferomedial means 'below and closer to the midline'.

Inter-: [L] indicates *between, among*. Intercourse is 'business between [individuals]' and may involve doing business or "doing the business"!; the interim is the 'between time'. The intertubercular groove of the humerus lies 'between the [greater and lesser] tubercles'; the interclavicular ligament runs 'between the clavicles'. Not to be confused with **intra-** (see below).

Intra-: [L] means *within, inside*. Something intramural lies 'within these walls'; intracellular means 'inside the cell'. Intraocular ('inside the eye') pressure increases in glaucoma. Not to be confused with **inter-** (see above).

Juxta-: [L] indicates *near, close to*. Juxtaposition involves 'placing things close to each other', often for the purpose of emphasising contrast. The juxtaglomerular apparatus of the kidney lies 'close to a glomerulus'. It comprises renin-secreting cells (lying between the afferent and efferent arterioles) and the macula densa cells (in the wall of the distal tubule).

Meso-: [Gr] indicates *middle, intermediate, between*. Mesopotamia was the land 'between the rivers' Tigris and Euphrates; in particle physics, a meson is a 'between particle' with a mass between an electron and a proton. The mesencephalon is the 'midbrain'; the mesoderm is the 'middle [germ] layer'. Note that meso- in Anatomy is also an abbreviation for mesentery, e.g. mesoappendix, mesocolon and mesosalpinx (mesenteries of appendix, colon and uterine tube respectively).

Meta-: [Gr] prefix with various meanings including *changing, transcending, going beyond*. Metabolism (the chemical 'change processes' of the body); metaphysics (something 'beyond physics'). The metaphysis of a bone is the 'growth change' region; the metacarpal bones are in the transition zone between the carpal (wrist) and phalangeal (finger) bones.

Multi-: [L] indicates *many, much*. Multiple, multiply, multi-dimensional, multi-ethnic and so on. The multifidus is a deep muscle of the back which is split into several parts.

Mylo-: [Gr] signifies *associated with the molar teeth* (**mylos**: [Gr] molar). On the medial aspect of the mandible, the mylohyoid line runs downwards and forwards from posterior molars and provides attachment for the mylohyoid m. The mylohyoid groove runs from the mandibular foramen to below the posterior molars and conveys the mylohyoid n to the muscle.

Myo-: [Gr] signifies *associated with muscle*. Myalgia ('muscle pain'), myology ('the study of muscle'), myocyte ('muscle cell'), myocardium ('muscle of the heart'), myosin (a muscle protein).

Neo-: [Gr] signifies *new*. A neonate is 'newborn'; a neologism is a newly-created word or expression; a neo-Nazi is one with extreme nationalistic or racist views.

-oid: [Gr] suffix meaning *of a certain shape or form* (**eidos**: [Gr]). Cricoid ('ring-shaped') cartilage, mastoid ('breast-shaped') process, pterygoid ('wing-shaped') bone, scaphoid ('shaped like the keel of a boat') fossa of sphenoid ('wedge-shaped or wedged-in') bone; sesamoid ('seed-like') bones.

Omo-: [Gr] refers to the *shoulder or scapula (= shoulder blade)*. The omohyoid muscle has attachments to the scapula (where exactly?) and to the hyoid bone.

Para-: [Gr] prefix meaning *near, beside, passing alongside*. Parallel ('alongside one another') lines; paramedic (works 'alongside a medic'). In Anatomy, paranasal air sinuses are 'beside the nasal [cavity]' and open into it; the parasympathetic nervous system operates 'alongside the sympathetic system'; the parathyroid glands lie 'near the thyroid gland'.

Peri-: [Gr] indicates *around, surrounding, enclosing*. Perimeter ('the around measurement'); periscope (allows you to 'see around'). Pericardium invests the heart; the peritoneum is 'stretched around' the abdomen and its viscera; periosteum 'invests bone'.

Post-: [L] signifies *behind, after, following*. Postgraduation is a stage 'after graduation'; a postscript is 'written after' the main body of a letter. Posterior 'behind' is from this root.

Pro-: [Gr,L] means *before in time or position, forward*. To prostrate ('lay down in front of'); prothrombin (the 'precursor' of thrombin); to produce ('bring forward'). Amongst anatomical terms with this word-root are protraction ('drawing forwards') and protrusion ('thrusting forward').

Quadr-, Quadri-: [L] indicates *four*. A quadrangle has 'four angles' and a quadrat has 'four sides'. The quadratus femoris m is shaped 'like a quadrat' whilst the quadriceps femoris m has 'four heads'; the quadrangular space of the arm has 'four angles' and transmits the posterior circumflex humeral vessels and axillary n (of course, you remembered!).

Retro-: [L] indicates *back, backwards, behind*. Retroactive ('coming into effect from a date in the past'); retrogressive ('backward walking', i.e. going back to an earlier, and by implication, an inferior, state). Retraction ('drawing backwards') of the tongue is effected by styloglossus and hyoglossus mm. Retraction of the upper limb is brought about by the latissimus dorsi and other muscles.

Semi-: [L] means *half*. A semicircle is 'half a circle'. The semilunar valves of the pulmonary artery and aorta are 'half-moon shaped'. How many valves are there in each of these arteries and how are they named?

Sub-: [L] denotes *under, beneath*. Sub-aqua sports involve going 'under water'; to subjugate is to 'put under the yoke'; subhuman is 'less than human'. The subarachnoid space lies 'below the arachnoid mater', between it and the pia mater; the sublingual gland lies 'beneath the tongue'; the subscapularis m is found 'under the scapula' when approached from posteriorly.

Super-, supra-: [L] means *above, over*. Superego (an 'above-normal ego'); supranational ('involving more than one nation'). Superior ('above') and suprarenal ('above the kidney') are anatomical terms from this same root.

Syn-, Sym-: [Gr] means *with, together*. Synchronise ('bring together in time'); synthesis ('bringing together to make a whole'); synergists 'work together' to produce a combined effect greater than the individual effects. What muscles act synergistically to cause flexion at the wrist joint? A symposium was originally a get-together not only for the purpose of discussion but also for drinking.

Tel-, Telo-: [Gr] means *an end*. From this root, we obtain words such as telomere ('the end part' of a chromosome), telophase ('the end phase' of cell division) and telencephalon ('the anterior end' of the forebrain, mainly comprising the cerebral hemispheres).

Tri-: [Gr,L] indicates *three, three times*. A triangle has 'three angles'; a tripod has 'three legs'. The trigone is a 'triangular' area of bladder between the openings of the ureters and urethra.

Uro-: [Gr] signifies *associated with urine or the urinary system*. Urea (a product of protein metabolism excreted in urine), ureter (the tube conveying urine from the kidney to the bladder), diuretic (an agent stimulating flow of urine), urinal (a place or receptacle for passing urine), urology ('the study of the [genito-]urinary tract').

Glossary 2

Some Useful Anatomical Terms, Meanings and Usages

Anatomical Names	Latin/Greek or Other Origins	Examples

A

Anatomical Names	Latin/Greek or Other Origins	Examples
Abdomen, abdominal abdominis	**Abdomen**: [L] *belly, paunch*; **abdominis**: [L] *of the abdomen.*	Abdomen
		Transversus abdominis m
Abduction, abducent, abductor	**Abducere**: [L] *to lead away*	Abduction
		Abducent n (cranial VI)
		Abductor pollicis longus m
Accessory	**Accedere**: [L] *to go/ come towards, to be added to*	Accessory nerve
Acetabulum, acetabular	**Acetum**: [L] *vinegar*, **acetabulum**: [L] *vinegar cruet, cup-shaped vessel*	Acetabulum, acetabular fossa
		Acetabular labrum
Acinus, acini, acinar	**Acinus**: [L] *berry, grape, pip*	Serous and mucous acini

Notes, Links and Non-Anatomical Usages

Anatomically, the abdomen is divisible into nine regions: superiorly are the epigastric and left and right hypochondriac regions; intermediate are the umbilical and left and right lumbar (lateral) regions; inferiorly are the hypogastric and left and right inguinal (iliac) regions. What bony and other landmarks are used to define the horizontal and vertical lines between these regions? Abdomen may derive from **abdere** [L] meaning to hide or retain. Certainly it hides and retains the abdominal viscera which may spill out when the anterior wall is cut

Literally, the transverse muscle of the abdomen. It is the deepest of the three anterolateral muscles (what are the other two?) and is supplied by ventral rami of T7-T12 and L1 spinal nn. It helps to control intra-abdominal pressure. What are its attachments and how does a knowledge of these attachments help to explain why the muscle does not contribute to the layers of fascia covering the testis and spermatic cord?

To abduct a person is to kidnap him/her (take him/her away by force or cunning). Abduction of the upper limb moves it away from the midline of the body. Brought about by deltoid m and what other muscle?

The link here is that the abducent n supplies the lateral rectus m which abducts (laterally rotates) the eyeball

The long abductor muscle of the thumb. Here, abduction is movement of the thumb to a position at which it is at right angles to the plane of the palm of the hand. What do you think the other thumb abductor m is called?

In a sense, the cranial part of the accessory n may be thought of as being accessory to the vagus n and its spinal part as adding to upper cervical spinal nn. What do these parts of cranial n XI supply?

Acetic acid (the main ingredient of vinegar!). The acetabulum or acetabular fossa, is the deep pelvic cup which houses much of the head of the femur. Presumably, it resembles an ancient vinegar cup!

The fibrocartilaginous rim of the cup attached to the acetabulum and the transverse acetabular lig

In parotid and pancreatic glands, the acini are arranged like bunches of grapes with the intercalated ducts being like the stalks

Anatomical Names	Latin/Greek or Other Origins	Examples
Acoustic	**Akouein**: [Gr] *to hear*	External acoustic meatus
Acromion, acromial, acromio	**Acros**: [Gr] *topmost, highest, outermost*; **omos**: [Gr] *shoulder*	Acromion
		Coracoacromial lig
		Acromioclavicular joint
Adduction, adductor	**Adducere**: [L] *to bring to, draw towards oneself*	Adduction
		Adductor brevis m
		Adductor pollicis m
Adenoid, adenoidal, adeno-	**Aden[as]**: [Gr] *gland*; **eidos**: [Gr] *shape, form*	Adenoids
		Adenohypophysis
Adipose, adipo-, adiposus	**Adeps**: [L] *fat*	Adipose tissue Adipocyte

Notes, Links and Non-Anatomical Usages

A canal leading inwards from the auricle (or pinna) of the external ear to the tympanic membrane. Does it follow a straight course? What is its nerve supply? Where is the internal acoustic meatus?

Acrobat (walker on tip-toe), Acropolis (High City), acrostic (lines of poetry or other writing whose first letters spell out a word or phrase). The first letters of the cranial nn spell out O,O,O,T,T,A, F,V,G,V,A,H which gives a useful mnemonic: On Old Olympus's Towering Top, A Fit Vice-Chancellor's Girlfriend Vaults and Hops! The acromion is a bony lateral projection of the scapular spine which forms the 'tip of the shoulder'

The ligament between the coracoid process and acromion forming an arch over the shoulder joint

A plane synovial joint between the lateral end of the clavicle and the medial side of the acromion. The joint houses an incomplete articular disc and is reinforced by acromioclavicular and coracoclavicular ligs

Adduce (to bring forth in evidence)

One of the adductors of the thigh at the hip joint. It is the smallest of the three muscles named adductor, the other two being? Adduction of the thigh involves moving it towards the midline

An adductor of the thumb. This movement brings the thumb parallel to the side of the palm and to the fingers. Are there any other thumb adductors?

Adenoids (literally, gland-shaped), adenoidal (having swollen adenoids), adenoma (a tumour of glandular tissue). Anatomically, the term adenoids is confined to the lymphoid glands of the nose and nasopharynx (otherwise known as the pharyngeal tonsils)

Part of the hypophysis (pituitary gland). It comprises the pars anterior (or pars glandularis!), pars intermedia and pars tuberalis

Adipose, adipocere (a waxy fat resulting from decomposition of the body under water). Adipose tissue is a connective tissue containing adipocytes (fat cells) specialized for fat storage. The two main types of adipose tissue are termed unilocular (white fat) and multilocular (brown fat). The two types differ in distribution and developmentally. How?

Anatomical Names	Latin/Greek or Other Origins	Examples
		Panniculus adiposus
Aditus	**Aditus**: [L] *access, entrance, opening*	Aditus of the mastoid antrum
		Aditus of the larynx
Afferent	**Afferens**: [L] *carrying towards*	Afferent nerve Afferent lymphatic
Ala, alae, alar	**Ala**: [L] *wing*	Alar ligs
Alveolus, alveoli, alveolar	**Alveolus**: [L] *bath, basin, small sac, cavity*	Dental alveolus
		Alveolar nn
		Lung alveoli
Ampulla, ampullae	**Ampulla**: [L] *bottle, narrow-necked flask*	Ampulla of the uterine tube
Amygdala, amygdaloid	**Amygdalum**: [Gr,L] *almond*; **eidos**: [Gr] *shape*	Amygdaloid body

Notes, Links and Non-Anatomical Usages

The fatty little web or network. Much white fat accumulates in the subcutaneous tissues or hypodermis where it is known as the panniculus adiposus. That of the abdomen may be several cm in thickness!

Adit (a term for an almost horizontal shaft into a mine). The aditus of the mastoid antrum connects the mastoid air cells to the tympanic (middle-ear) cavity

Otherwise known as the laryngeal inlet. Its borders are anterosuperior (the epiglottis), lateral (the aryepiglottic folds) and posteroinferior (the interarytenoid mucosal fold)

In Anatomy, afferent usually implies conveying neural information towards the CNS but is also used to describe blood or lymphatic vessels supplying an organ, gland or other structure, e.g. afferent lymphatics to lymph node, afferent arteriole to renal glomerulus

The two alar ligs are strong ligaments which run like wings from the dens of the axis to the medial borders of the occipital condyles. They prevent excessive rotation at the atlantoaxial joints. Various other anatomical structures have a wing or ala, e.g. the cerebellum, crista galli of the ethmoid bone, ilium, sacrum, vomer

The dental alveolus is the tooth socket

These nerves supply teeth by entering their roots via their alveoli

In section, the lung looks like a sponge – full of little spaces!

Ampullary. The ampulla of the uterine tube accounts for over half of total tube length and lies between the infundibulum and isthmus. It is thin-walled with a much-folded mucosa. Other structures also have ampullae, e.g. the ductus deferens and semicircular canals

Amygdalin (bitter-tasting extract from almonds). The amygdaloid body is in the walls of the inferior horn of the lateral ventricle in the temporal lobe of the cerebrum. It is associated with the sense of smell (olfaction) so should be particularly good at detecting cyanide which is found in bitter almonds!

Anatomical Names	Latin/Greek or Other Origins	Examples
Anastomosis, anastomotic	**Ana-**: [Gr] *up, build up*; **stoma**: [Gr] *mouth*	Arterial anastomosis
Anconeus	**Ancon**: [L] *elbow*	Anconeus m
An[n]ulus, annular	**Anulus**: [L] *little ring*	Anulus fibrosus
		Annular lig
Ansa	**Ansa**: [L] *loop, handle* (as on a jug)	Ansa cervicalis
Antagonist	**Ant-**: [Gr] *against*; **agon**: [Gr] *contest, struggle* **agonistes**: [Gr] *rival, competitor*	Antagonistic muscle
Antrum, antra, antral	**Antrum**: [L] *cave, cavern, hollow*	Pyloric antrum
		Mastoid antrum

Notes, Links and Non-Anatomical Usages

Anastomosis (equipped with a mouth or junction), anastomotic. Anastomoses are circulatory safety devices to ensure continuity of flow to certain areas (e.g. arterial anastomoses) or devices for by-passing flow to certain regions (e.g. arteriovenous shunts). Arteries which do not anastomose are called end arteries. What is the difference between an anatomical end artery and a physiological end artery? Can you think of examples of each?

Ancon (an architectural term for a bracket). The anconeus m (nerve supply?) is an extensor of the elbow

Annulate. The an[n]ulus fibrosus (fibrous ring) is an important component of the intervertebral disc (the other being the nucleus pulposus which it encircles and contains). The anulus has outer collagenous and inner fibrocartilaginous zones. With advancing age, the anulus may weaken and the nucleus burst through, usually in a posterolateral direction. This happens in lumbago and sciatica

The annular lig of the proximal radio-ulnar joint is a strong band which encircles the head of the radius keeping it against the radial notch of the ulna. Within the ligament, the radius is free to rotate. During which forearm movements does rotation occur?

The ansa cervicalis (loop of the neck) is formed by the union of an upper (spinal C1 and cranial XII nerves) and a lower (C2, C3 nerves) root which form a loop in front of the carotid sheath. It lies in which triangle of the neck? Which muscles does it supply?

Antagonism, antagonize, agony, agonist. Here, an antagonist is a muscle or muscle set which opposes the action(s) of another muscle or set. Antagonistic relationships may vary with the joint movement. Thus, in wrist flexion-extension, the flexors and extensors are antagonists but in adduction (medial or ulnar deviation) and abduction (lateral or radial deviation), some of the flexors are antagonists (e.g. flexor carpi ulnaris m versus flexor carpi radialis longus and brevis mm)

The pyloric antrum is caudal to the angular incisure of the stomach and leads into the pyloric canal which ends at the pylorus (opening into the duodenum)

An air sinus in the mastoid process of the temporal bone whose relations are surgically important because it can be infected. What are its anterior, posterior, superior, inferior, medial and lateral relations? It opens into the tympani cavity via the aditus of the antrum

Anatomical Names	Latin/Greek or Other Origins	Examples
Anus, anal, ani	**Anus**: [L] *ring, rectum*	Anus
		Anal sphincter
		Levator ani m
Aponeurosis	**Apo-**: *from, off;* **neuro**: [Gr] *nerve, muscle*	Palmar aponeurosis
Appendix, appendicular	**Appendix**: [L] *appendage*	Vermiform appendix
		Appendicular a
Aqueous	**Aqua**: [L] *water*	Aqueous humour
Arachnoid	**Arachne**: [Gr] *spider, cobweb;* **eidos**: [Gr] *shape*	Arachnoid mater

Notes, Links and Non-Anatomical Usages

Not to be confused with **anus**: [L] *old woman*, or **annus**: [L] *year* (although the Queen's annus horribilis may have caused this senior citizen a pain in the backside!). Anulus (little ring) comes from this same word-root. People sometimes refer to their anus as 'my ring'

Has internal and external components. The internal sphincter is involuntary. The external is voluntary and has three parts: subcutaneous, superficial and deep. The internal sphincter is supplied by sympathetic nn from the inferior hypogastric plexus. What is the innervation of the external sphincter?

The main part of the pelvic diaphragm dividing the pelvis from the perineum. What is the other muscle of the diaphragm? The levator ani mm form a sling to support pelvic viscera and are supplied by the perineal branches of S4 and the pudendal n or the inferior rectal n

An aponeurosis is a flat tendon arising from a muscle. In the case of the palmar aponeurosis, the muscle is palmaris longus which is absent on one or both sides in about 10% of people. Do you have a palmaris longus m? How would you find out?

Append, appendage. If the vermiform (worm-like) appendix is inflammed (appendicitis), it may need to be removed surgically (appendicectomy). But sometimes it is hidden from view! So what useful features of the colon might the surgeon follow to locate it?

Supplies the appendix as a branch of the posterior caecal a from the ileocolic a. What are the routes of venous and lymphatic drainage of the appendix?

Aqueous, aquatic, aqua vitae (water of life – a name given to intoxicating liquors in many languages, e.g. whisky derives from **uisge-beatha**: [Gael] meaning water of life), aqueduct (water-channel – e.g. cerebral aqueduct). The aqueous humour fills the anterior and posterior chambers of the eye. It is produced by capillaries in the ciliary processs and drains away via anterior ciliary vv. Imbalances of production versus resorption may increase intraocular pressure. This condition is called what?

Arachnids (the class of animals that includes spiders and scorpions), arachnophobia (fear of spiders). The arachnoid mater is one of the three meninges covering the brain and spinal cord. Sandwiched between the outer dura mater and inner pia mater, it forms a delicate membrane deep to which is the subarachnoid space criss-crossed by a fine network of threads which resembles a spider's web!

Anatomical Names	Latin/Greek or Other Origins	Examples
Arcuate	**Arcus**: [L] *arch, bow, curve;* **arcuatus**: [L] *arched*	Arcuate ligs
		Arcuate line
Areola, areolae, areolar	**Areola**: [L] *small open space*	Areolar tissue
		Areola of nipple
Articulation, articular	**Articulare**:[L] to *divide into joints, to speak distinctly;* **articulatio**: [L] *a joint*	Articulation
		Articular cartilage
Arytenoid, ary-	**Arytena**: [Gr] *funnel*; **eidos**: [Gr] *shape*	Arytenoid cartilages
		Aryepiglottic folds
Atlas, atlanto-	**Atlas**: [Gr] *a mythical Greek giant*	The atlas
		Atlanto-occipital joint
Atrium, atria, atrial, atrio-	**Atrium**: [L] *part of house next to entrance, hall* (i.e. where visitors are greeted)	Left and right atrium

Notes, Links and Non-Anatomical Usages

The five curved arcuate ligs (1 median, 2 medial, 2 lateral) attach the diaphragm to the posterior abdominal wall

The arched line at which the anterior and posterior layers of the rectus sheath alter in composition, midway between umbilicus and pubic symphysis. Also, the iliac part of the linea terminalis which divides the greater from the lesser pelvis

Areolar connective tissue is sometimes called loose connective tissue because of the open spaces seen on light microscopic examination

The pigmented area of skin around the nipple which becomes larger and darker in females during pregnancy. Subareolar glands secrete a protective lubricant during lactation

Article (a distinct item), articulate (able to speak well), articulation (synonymous with joint)

Found on the articular surfaces of bony elements in cartilaginous joints

This is a mystery to me. The cartilages are not funnel-shaped but more like triangular pyramids. Perhaps the funnel, here, refers to the wide laryngeal opening (on the posterior summit of which the arytenoid cartilages sit) leading to the narrower tracheal lumen

Mucosal folds running from the arytenoids to the epiglottis and containing muscle fibres (aryepiglottic mm, continuations laterally of the oblique arytenoid mm) and minor cartilages of the larynx

The giant Atlas carried the World on his shoulders. The first cervical vertebra (the atlas) carries the head

The joint between the superior articular facets of the atlas and the occipital condyles. The joint allows flexion-extension of the skull on the atlas with some lateral flexion. We nod our heads at atlanto-occipital joints but shake our heads at which other joints?

The atria are the receiving chambers of the heart!

Anatomical Names	Latin/Greek or Other Origins	Examples
		Atrioventricular valves
Audition, auditory	**Audire**: [L] *to hear*	Audition
		Auditory tube
		Auditory ossicles
Auricle, auricular[is], auriculo-	**Auris**: [L] *ear;* **auricula**: [L] *little ear, lobe of ear*	Auricle
		Auriculotemporal n
		Auricularis mm
Auscultation	**Auscultare**: [L] *to listen to*	Auscultation
Axilla, axillae, axillary	**Axilla**: [L] *arm-pit*	Axilla
		Axillary n
		Axillary tail

Notes, Links and Non-Anatomical Usages

Lie between atrium and ventricle. That on the left has two cusps and that on the right has three. Coincidentally, the left lung has two lobes and the right has three

Audible, audience, audition (the special sense of hearing), auditorium, audiovisual (involving sound and sight)

Joins the middle ear cavity to the nasopharynx, hence its alternative title - the pharyngotympanic tube

Three tiny bones (ossicle means little bone) which run across the tympanic cavity from the tympanic membrane to the fenestra vestibuli. What are the names of the three bones?

Aural. The auricular appendages of the heart atria resemble little ears

Sensory branch of the mandibular division of the trigeminal (cranial V) n supplying the skin of the ear and temple. Postganglionic (from which ganglion?) parasympathetic fibres of the glossopharyngeal (cranial IX) n hitch a ride on this nerve and leave it to provide secretomotor innervation to the parotid gland

A set of very minor muscles which move the auricle of the ear and are supplied by the facial (cranial VII) n. They are often vestigial

A form of examination involving listening to body sounds, usually with the aid of a stethoscope!

The axilla is a pyramidal region between the upper thoracic wall and arm. The anterior wall is formed by the pectoral mm, the posterior by subscapularis, teres major and latissimus dorsi mm and the medial by ribs 1-4 and their intercostal spaces and serratus anterior m. What structures form the lateral wall?

Provides the motor supply to the deltoid and teres minor mm. What is its cutaneous supply?

Runs from the superolateral quadrant of the breast along the lower border of pectoralis major m and towards the axilla and its pectoral group of lymph nodes

Anatomical Names	Latin/Greek or Other Origins	Examples
Axis, axial	**Axis**: [L] *axle, pivot*	Axis

B

Anatomical Names	Latin/Greek or Other Origins	Examples
Base, basal, basilar, basilic	**Basis**: [L] *base, pedestal*	Base of the heart
		Basal nuclei or ganglia
		Basilar a
		Basilic v
Biceps, bicipital	**Biceps**: [L] *two-headed*	Biceps brachii m
		Bicipital groove
Bicuspid	**Bicuspid**: [L] *having two points*	Bicuspid valve
Bifurcation	**Bifurcus**: [L] *two-forked;* **furca**: [L] *fork*	Bifurcation of the abdominal aorta
Brachium, brachial, brachio-	**Brachium**: [L] *arm*	Brachium
		Antebrachium

Notes, Links and Non-Anatomical Usages

The axis is the 2nd cervical vertebra. It bears a prominent anterior peg called the dens or odontoid process. Rotation of the atlas and skull around the odontoid process occurs at the atlanto-axial joint

The base of the heart lies posteriorly where great vessels leave (aorta, pulmonary trunk) or enter (SVC and IVC). Which heart chamber mostly forms the base? It is worth noting that the base or fundus of an organ is often situated furthest away from the anterior surgical approach

A collection of various large subcortical nuclei situated at the base of the forebrain

Lies at the base of the skull and is formed by the union of the two vertebral aa. It ends by dividing into two posterior cerebral aa. What are its branches and distributions?

Drains the lower (more distal) and medial part of the upper limb

Biceps brachii m (two-headed muscle of the arm) is a major flexor of the elbow joint and a powerful supinator when the elbow is flexed. Its long head runs in the intertubercular groove of the humerus and over the shoulder joint to the supraglenoid tubercle of the scapula. To which bit of the scapula does the short head attach?

Another name for the intertubercular groove

The modern definition of bicuspid is having two cusps (rather than having two points). The bicuspid valve is the left atrioventricular or mitral valve. The premolar teeth are also bicuspid

Furcula (a forked structure such as the wishbone of a bird or the tail pincers of the earwig, otherwise known as the forky tail!), bifurcate (divide into two forks or branches). The abdominal aorta bifurcates at the level of the 4th lumbar vertebra into the left and right common iliac aa

Brachiate (swing from tree to tree using arms), Brachiosaurus (arm-lizard) was a dinosaur with forelimbs much longer than its hindlimbs. Anatomically, the arm extends from shoulder to elbow

The forearm. It extends from elbow to wrist

Anatomical Names	Latin/Greek or Other Origins	Examples
		Brachial a
		Brachialis m
		Brachioradialis m
Bregma	Uncertain. Maybe from **bregmenos**: [Gr] *wet, moist*	Bregma
Brevis	**Brevis**: [L] *short, brief*	Adductor brevis m
Bucca, buccae, buccal, bucco-	**Bucca**: [L] *cheek, mouth*	Buccal n
		Buccopharyngeal fascia
Buccinator	**Buc[c]inator:** [L] *trumpeter*	Buccinator m
Bulb, bulbo-	**Bolbos**: [L] via [Gr] *onion, bulbous root*	Olfactory bulb

Notes, Links and Non-Anatomical Usages

Arises from the axillary a as it enters the arm. It can be palpated medially as it lies on the brachialis m and is overlapped laterally by the biceps brachii m. To feel the pulse, push the latter muscle laterally and press laterally (not more deeply)

Arises from the antero-inferior aspect of the humerus and inserts on the coronoid process of the ulna. It is supplied by the musculocutaneous n with two other muscles (why is it useful to remember BBC for the names of these three muscles?)

A flexor of the elbow and rotator of the forearm running from the lateral supracondylar ridge of the humerus to the lower lateral aspect of the radius. What is its nerve supply? In what position of the forearm does it act most effectively as an elbow flexor?

The area of the skull where the sagittal and coronal sutures meet. It may derive its name from the fact that this area feels soft and moist in the neonate because the frontal and parietal bones are unfused and surround the anterior fontanelle

Brief, brevity. If a muscle is called X brevis, this implies that there is at least one other muscle called X longus. In the case of the femoral adductors, there is also a third, viz. adductor magnus m. Adductor brevis is the shortest of the three femoral adductor mm. Isn't it funny that abbreviate (to shorten) is such a long word!

Buccal (concerning or related to the cheek). The buccal n is sensory to what areas of skin and mucosa?

The tough fascia covering the external surface of the pharynx

Watch the cheeks of a trumpeter at full blow – the rounded cheek contours are contained by distended buccinator mm. What are the nerve supply and normal actions of this muscle?

Bulbous (bulging, fat, round). The olfactory bulb is an oval mass distending the olfactory tract where it lies on the cribriform plate of the ethmoid bone. Nerve fibres from cells in the olfactory mucosa pass through the foramina in the plate and enter the bulb

Notes, Links and Non-Anatomical Usages

This lies between the crura of the penis and is attached to the inferior aspect of the perineal membrane. Anteriorly, it narrows and runs into the corpus spongiosum. It lies between the membranous and penile portions of the urethra and is vulnerable to injury by impact with horizontal beams, bicycle crossbars or similar. If it ruptures, what path would leaking urine take?

In males, the muscle helps to empty the urethra during urination and ejaculation. In females, it constricts the vaginal orifice and contributes to erection of the clitoris. So, remember it as 'the oomph and tweak muscle'!

Bull (as in 'papal bull' and from the same word-root as bulletin), bullet. The ethmoidal bulla is a promontory on the lateral wall of the nasal cavity (middle meatus) caused by the middle ethmoidal air sinuses

Bursitis, bursar (treasurer of a college or university – holding on to the sack of money?), bursary (scholarship award – usually not a sack of money!). I have suffered from inflammation of the subacromial bursa (subacromial bursitis). Where do you think the bursa lies? Is it connected to the shoulder joint cavity? What are the effects of its bursitis?

Caecum (short for 'intestinum caecum' or blind ending of the intestine). Indeed, it is a cul-de-sac from which the appendix opens

There is more than one! That of the tongue is a shallow pit in the middle of the sulcus terminalis on its dorsal surface. It is an embryological remnant of the thyroglossal duct which formed during descent of the thyroid gland. The other foramen caecum is not blind but transmits an emissary vein which can carry infection from the nose to the brain and meninges. Where is it?

Calcaneus – the heel bone. The Greek hero, Achilles, was fated to die young so his mother tried to make him immortal by dipping him in the River Styx. However, she held him by his heel and this became a vulnerable spot where an arrow eventually hit and killed him. Hence the phrase, Achilles heel, referring to a person's weak spot!

The tendo calcaneus (calcaneal or Achilles tendon) is the powerful tendon into which the gastrocnemius and soleus mm run. It inserts on the posterior aspect of the calcaneus

Anatomical Names	Latin/Greek or Other Origins	Examples
		Calcaneonavicular lig
Calvaria	**Calva**: [L] *skull*	Calvaria
Calyx, calyces, calycal	**Kalyx**: [Gr] *cup, case of a bud*	Renal calyx
Cancellous	**Cancelli**: [L] *crossbars, lattice, railings*	Cancellous bone
Canine	**Canis**: [L] *dog, dog-like*	Canine tooth
Canthus, canthi	**Kanthos**: [Gr] *angle of the eye*	Canthus
Capitulum	**Capitulum**: [L] *small head*	Capitulum
Caput, capitis, capitate	**Caput**: [L] *head*	Caput Medusae

Notes, Links and Non-Anatomical Usages

The plantar calcaneonavicular (or spring) lig is very strong and rather elastic. It runs from the anterior of the sustentaculum tali of the calcaneus to the inferior of the navicular tuberosity and is the plantar ligament of the talocalcaneonavicular joint. It helps to maintain the medial longitudinal arch of the foot

Calvary (the hill where Jesus Christ was crucified) comes from the same word-root since it is a translation, via Greek (**Golgotha**: [Gr] *place of skulls*), of the Aramaic word Gulgulta The term calvaria is now restricted to describe the skull cap. Calvities is a term used to describe baldness affecting the top of the head. Interestingly, **calvo** [Sp] means bald!

The minor calyces of the kidney are cup-like extensions of the pelvis into which the papillary ducts of nephrons open. Minor calyces unite to form major calyces which converge on the renal pelvis proper. From this, the ureter arises

Cancel (to cross out, to put a cross through), cancellation. This type of bone is otherwise known as spongy or trabecular bone. It occurs in vertebral bodies and towards the ends of long bones. It is characterised by struts or trabeculae which surround pores containing bone marrow. It is less dense and weaker than compact or cortical bone

This is a pointed tooth lying between the incisors and premolars. It is sometimes called a dog-tooth because it resembles the equivalent (but much larger) tooth in dogs. It is specialised for gripping food. By the way, the Canary Isles were so named by the Romans because they were home to large dogs and not canaries!

Where the upper and lower eyelids meet at the corners of the eye are known as the medial and lateral canthi

Capitulate (to draw up terms of surrender under various headings). The rounded capitulum at the distal end of the humerus articulates with the shallow fossa of the head of the radius

Capital (main city), capitation (a poll tax or tax per head), captain (head person on board ship), decapitation ("Off with his head!"). One of the visible signs of portal hypertension is a set of swollen veins radiating from the umbilicus like the 'head of the Medusa' (caput Medusae), a mythical creature who had hissing snakes instead of hair!

35

Anatomical Names	Latin/Greek or Other Origins	Examples
		Splenius capitis m
		Capitate bone
Cardiac, cardio-	**Cardia**: [Gr] *heart;* **cardiacus**: [L] *relating to heart, or to upper part of stomach where the oesophagus enters*	Cardiac vv
		Cardiac notch of the lung
Carina	**Carina**: [L] *keel, ship*	Carina tracheae
Carotid	**Karotis**: [Gr] *drowsiness*	Carotid a
		Carotid bifurcation
Carpus, carpal, carpi, carpo-	**Karpos**: [Gr] *wrist;* **carpere**: [L] *pluck, seize*	Carpal bones

Notes, Links and Non-Anatomical Usages

An extensor and rotator of the head running from the mastoid process and superior nuchal line to the ligamentum nuchae and spinous processes of vertebrae C7-T3. It is supplied by dorsal rami of middle cervical spinal nn. It gets its name (bandage of the head) from its long, thin, strip-like appearance

The capitate is the largest and, therefore, the head carpal bone

Cardiology. The term cardia is used to refer to the oesophageal opening of the stomach. Cardiac vv drain the heart and (except for which veins?) mostly open into the right atrium via the coronary sinus

Owing to the leftward disposition of the heart and pericardium, the anterior border of the left lung deviates to the left below the 4th costal cartilage and then curves down towards the 6th

Carinate (having a keel). The carina of the trachea is the keel-like inferior part of the last (lowest) tracheal cartilage and runs in the gap created by the tracheal bifurcation into main bronchi. The term is sometimes used to refer to the bifurcation itself (vertebral level T4-T5)

The arteries were so-named from the belief that their compression caused drowsiness or stupor. The left common carotid a arises from the aortic arch. The right common carotid a arises where?

At the level of the upper border of the thyroid cartilage (corresponding to a horizontal plane passing through the disc between C3 and C4 vertebrae), the common carotid aa bifurcate into the internal and external carotid aa. The cervical part of the internal carotid a has no branches. But what are the branches of the external carotid a?

Carpe diem (seize the day! – to grip something tightly, you need to extend at the wrist). Carpus is the anatomical name for the wrist. What are the names of the carpal bones?

Anatomical Names	Latin/Greek or Other Origins	Examples
		Flexor carpi ulnaris m
		Carpometacarpal joints
Caruncle, caruncular	**Caruncula**: [L] *a small morsel of meat* (hence, a small fleshy eminence)	Lacrimal caruncle
Cauda, caudal, caudate	**Cauda**: [L] *tail*	Cauda equina
		Caudate lobe
		Caudate nucleus
Cava, cavae, caval	**Cavus**: [L] *hollow*	Venae cavae
		Caval opening in the diaphragm

Notes, Links and Non-Anatomical Usages

Flexor of the wrist on the ulnar (medial) side. Supplied by ulnar n

Except for the first (that of the thumb), these are plane synovial joints allowing flexion, extension, adduction, abduction and some rotation. That of the thumb is a synovial saddle joint which also allows circumduction

Caruncle (the fleshy outgrowth which is the cock's comb). The lacrimal caruncle is a reddish fleshy mass of skin at the medial angle of the eye and contains sebaceous and sweat glands

Caudal implies closer to the tail. The cauda equina (horse's tail) arises because the spinal cord terminates in the upper lumbar region but lower spinal n roots have to travel progressively longer distances to their vertebral exits. The appearance in the vertebral canal is of a horse's tail!

A small lobe at the posterior of the visceral surface of the liver and sandwiched between the IVC and lesser omentum

Together with the lentiform (lens-shaped) nucleus, it forms the corpus striatum (striated body) of the forebrain. The caudate (tail-bearing) nucleus has a head, body and tail

Nothing to do with cava, the Spanish sparkling wine! The superior and inferior venae cavae drain into which heart chamber?

Several structures pass between the thorax and abdomen by penetrating it or passing around it. Amongst the openings in the diaphragm are those for the IVC (level of 8th thoracic vertebra and, coincidentally, vena cava has eight letters) and oesophagus (level of 10th thoracic vertebra and having 10 letters). The aorta passes posterior to the diaphragm via the aortic hiatus (12 letters and guess what vertebral level?)

Anatomical Names	Latin/Greek or Other Origins	Examples
Cavernous	**Caverna**: [L] *cave, hollow, cavern*	Cavernous sinus
		Corpus cavernosum
Cephalon, cephalic	**Cephalon**: [Gr] *head*	Encephalon
		Mesencephalon
		Cephalic v
Cerebellum, cerebellar	**Cerebellum**: [L] *small brain*	Cerebellum
		Cerebellar peduncles
		Tentorium cerebelli
Cerebrum, cerebral, cerebro-	**Cerebrum**:[L] *brain*	Cerebrum
		Cerebral cortex

Notes, Links and Non-Anatomical Usages

Cave, cavern (a large cave). A cavernous venous sinus lies on each side of the body of the sphenoid bone and extends from the apex of the petrous part of the temporal bone to the superior orbital fissure. The nerves associated with its lateral wall are "O, Tom!" being, from above downwards, the Oculomotor (cranial III), Trochlear (IV), Ophthalmic (V) and Maxillary (V) nn. What other important structures are associated with the sinus?

The paired corpora cavernosa are masses of erectile tissue that account for much of the volume of the penis and clitoris. Do they lie ventrally or dorsally? So...is the anatomical position of the penis, flaccid or erect?

Cephalopod (foot-headed creature, like an octopus or squid, in which the feet or tentacles look as if they grow out of the head), Bucephalus (Oxhead - the favourite horse of Alexander the Great - had a white, ox-like patch on its head), encephalogram (an image of the brain), hydrocephalus ('water on the brain' – cured by a tap on the head?). The brain (encephalon) is contained in the head

The midbrain connects the pons and cerebellum (hindbrain or rhombencephalic structures) with the forebrain (prosencephalon)

Drains at the head-end of the upper limb rather than the bottom-end (basilic v). Into which vein does the cephalic v drain?

Cervelat and saveloy sausages both come from this root and, presumably, once contained brain!

Stalks of projection fibres connecting the cerebellum to the midbrain (superior cerebellar peduncles), pons (middle) and medulla oblongata (inferior cerebellar peduncles) and, hence, to other sites

The tent of the cerebellum is made of dura mater and covers the cerebellum in the posterior cranial fossa. It is attached to the occipital (posteriorly), temporal (laterally) and sphenoid (anteriorly) bones

Cerebration (using the brain), decerebrate (having no brain, usually because it has been removed!)

The human cerebral cortex is gyrencephalic (complexly folded into gyri and sulci). Although its surface is very extensive, about two-thirds of the total is hidden from view within the sulci and the insula. By contrast, the brain of a mouse is lissencephalic (rather smooth and lacking complex gyri and sulci)

Anatomical Names	Latin/Greek or Other Origins	Examples
Cervix, cervical[is]	**Cervix**: [L] *neck*	Uterine cervix
		Cervical vertebra
Chiasma, chiasmata	**Chiasma**: [Gr] *crossing-over*	Optic chiasma
Choana, choanae	**Choane**: [Gr] *funnel*	Choana
Chondral, chondro-	**Chondros**: [Gr] *cartilage, gristle*	Costochondral joint
		Chondrocranium
Chorda, chordae	**Chorda**: [L] *cord, rope*	Chorda tympani
		Chordae tendineae
Choroid	**Chorion**: [Gr] *chorion*; **eidos**: [Gr] *form*	Choroid (eyeball), choroid plexus (ventricles)
Ciliary	**Cilium**: [L] *eye, eyelash*; **supercilium**: [L] *eyebrow*	Ciliary body, ganglion, etc

Notes, Links and Non-Anatomical Usages

The uterine cervix projects as a knob into the vagina and is surrounded by the fornix which is divided into anterior, posterior and lateral fornices. What pelvic structures can be palpated by inserting a finger into these fornices?

How many cervical vertebrae are there? How many does a giraffe have?

Ultimately stemming from resemblance to the Greek letter chi which is cross-shaped. Exactly the shape of the optic chiasma at which fibres from the nasal (medial) halves of the retina become contralateral whereas those from the temporal (lateral) halves remain ipsilateral. During meiosis, maternal and paternal chromatids swap genetic material in the process called chiasma formation or crossing-over!

The nasal choanae (posterior nasal apertures) funnel inspired air into the nasopharynx

Chondrocyte, chondroblast, achondroplasia, chondrichthyes (the cartilaginous fish, including the sharks and rays). Costochondral joints occur between ribs and their cartilages. There is no movement between them since the cartilage is merely the unossified remnant of rib development from cartilage

Parts of the skull developing or remaining in a cartilaginous state

The chorda tympani (cord of the eardrum) is a branch of the facial (cranial VII) n which leaves the skull at the petrotympanic fissure and joins the lingual n. In part of its course, it runs across the medial side of the tympanic membrane

Tendinous cords passing from a given papillary mm to adjacent sides of an adjacent pair of cusps of an atrioventricular valve

Choroid, at whatever site, resembles chorion in containing many blood vessels

Cilia resemble tiny eyelashes, a supercilious person looks at you with raised eyebrows (and looks down the nose!). The ciliary body, ganglion and muscles are all associated with the eye

Anatomical Names	Latin/Greek or Other Origins	Examples
Cingulate	**Cingula**: [L] *belt, girdle*	Cingulate gyrus
Circumflex	**Circum**: [L] *around, about;* **flectere**: [L] *to bend*	Circumflex artery
Circumvallate	**Circum**: [L] *around, about;* **vallum**: [L] *rampart*	Circumvallate or vallate papillae
Cisterna, cisternae, cisternal	**Cisterna**: [L] *reservoir, cistern*	Cisterna magna
		Cisterna chyli
Claustrum	**Claustrum**: [L] *bolt, bar, lock, enclosed space*	Claustrum
Clavicle, clavicular, clavi-, -clavian, -clavius	**Clavis**: [L] *key, bolt, bar*	Clavicle
		Clavipectoral fascia
		Subclavian aa
		Subclavius m

Notes, Links and Non-Anatomical Usages

Surcingle (a strap running round a horse to keep a rug in place), cinch (a girth for a Western saddle) comes from the same word-root via **cincha**: [Sp] *girth*. The cingulate gyrus lies on the medial aspect of the cerebrum and, with the cingulate sulcus, partly encircles the corpus callosum

Circumflex implies curving or bending around and aptly describes the coronary, femoral, fibular, humeral, iliac and scapular circumflex aa

Large taste buds found running parallel and anterior to the sulcus terminalis on the dorsum of the tongue. Embryologically and histologically, they belong to the posterior third of the tongue. So what is their innervation?

Cistern. The cisterna magna (big cisterna), otherwise known as the cerebello-medullary cistern, is a subarachnoid space containing cerebrospinal fluid

This 'juicy cistern' is a narrow lymphatic (lymph is the juice) sac running posterior to the abdominal aorta and IVC and becoming continuous with the thoracic duct. Intestinal and lumbar lymphatic trunks from the intestines, lower limbs and lower trunk drain into it

Claustrophobia (fear of enclosed spaces), cloister (a covered walkway in a religious building). The claustrum is an area of brain matter sandwiched between the outer insular cortex and the deeper external capsule and lentiform nucleus

The collar-bone is the clavicle (little key - named because its shape resembles that of a key used by the ancients), clavichord (a keyboard instrument)

Deep fascia ensheathing the pectoralis minor and subclavius mm and attached to the clavicle and axillary fascia

These arteries pass 'below the clavicle' and become the axillary aa as they pass into the axilla. The branches of the subclavian a are the vertebral a, thyrocervical trunk, costocervical trunk and internal thoracic a. What areas do they supply?

A small muscle running from the 1st rib and costal cartilage to the inferior surface of the middle third of the clavicle. The muscle is supplied by spinal nn C5 and C6 but an important function for the muscle is hard to find. It presumably depresses the lateral end of the clavicle

45

Anatomical Names	Latin/Greek or Other Origins	Examples
Cleido-	**Cleido**-: [Gr] *relating to the clavicle*	Sternocleidomastoid m
Clinoid	**Klinein**: [Gr] *to slope, lean*	Clinoid processes
Clivus	**Clivus**: [L] *ascent, hill*	Clivus
Coccyx, coccygeal	**Coccyx**: [L] *cuckoo*	Coccyx
		Anococcygeal body
		Coccygeus m
Cochlea, cochleae, cochlear	**Cochlea**: [L] *snail, snail shell*	Cochlea
		Vestibulocochlear n (cranial VIII)
Coeliac	**Coeliacus**: [L] *related to the stomach*; **Koilia**: [Gr] *stomach, belly, abdomen*	Coeliac plexus
Coelom, coelomic	**Koilos**: [Gr] *hollow*; **koiloma** [Gr] *cavity*	Coelom (or coelomic cavity)
Coeruleus	**Caeruleus, coeruleus**: [L] *sky-blue, blue*	Locus coeruleus or caeruleus

Notes, Links and Non-Anatomical Usages

This muscle attaches to the mastoid process and the sternum and clavicle. It is supplied by the accessory n. How would you test the muscle's action in a patient?

Incline, inclination. The posterior margins of the lesser wing of the sphenoid bone end medially in the anterior and posterior clinoid processes. Between all four processes is the sella turcica (Turkish saddle) which forms a depression, the hypophyseal or pituitary fossa

The clivus is the slope (made by the sphenoid and occipital bones) extending upwards and forwards from the foramen magnum. In the midline, it forms an anterior boundary of the posterior cranial fossa

In sagittal section, the coccyx is shaped like a cuckoo's beak

A midline raphe for part of the levator ani m and lying between the coccyx and anus

The posterolateral component of the pelvic diaphragm attached to the ischial spine and sacrococcyx and supplied by spinal nn S4 and S5

The cochlea of the inner ear resembles a spiral snail shell

This nerve is associated with balance (equilibration) and hearing (audition). What artery accompanies the nerve into the internal acoustic meatus?

The coeliac plexus is the largest autonomic plexus found posterior to the stomach and surrounding the roots of the coeliac and superior mesenteric aa. Coeliac disease (a gluten hypersensitivity affecting the small intestine rather than the stomach)

The mesodermal body cavity containing abdominal and other viscera. The coelacanth (hollow spine) is a fish, once thought extinct, which has fins bearing hollow spines

Cerulean (a deep blue, like the sky). The locus (place, spot) coeruleus lies in the dorsal wall of the rostral pons in the floor of the 4th ventricle and appears bluish in colour in unstained sections. It is concerned with responses to stress/panic and is highly active in noradrenaline production. Those with Alzheimer's disease may suffer loss of a substantial number of neurons in this region

Anatomical Names	Latin/Greek or Other Origins	Examples
Collateral	**Col-**: [L] *together with*; **latus**: [L] *side*	Fibular collateral lig
		Collateral circulation
Colliculus, colliculi	**Colliculus**: [L] *little hill*	Superior colliculus
Colon, colic, colo-	**Colon**: [L] *large intestine*	Colon
		Colic aa
Comitans, comitantes	**Comes**: [L] *companion, attendant, partner*	Venae comitantes
Commissure, commissural	**Commissura**: [L] *joint, connection*	Anterior commissure
		Commissural fibres

Notes, Links and Non-Anatomical Usages

Collateral here has the sense of supportive. The fibular collateral lig is found on the fibular (lateral) side of the knee joint. Is this ligament attached to the lateral meniscus?

When the blood supply to a particular region is interrupted, compensatory circulations may open up. For instance, superior and inferior ulnar collateral aa arise from the brachial a and anastomose around the elbow joint with ulnar recurrent aa from the ulna a. The clinical importance of these collaterals becomes apparent when the brachial or ulna aa are obstructed near the elbow

The superior and inferior colliculi together make up the corpora quadrigemina (the quadruplet bodies) and sit on the tectum of the midbrain

Colonic irrigation (more a fashion statement than a therapy), colostomy (a surgical intervention to create an artificial opening from the colon to serve as an anus). The colon is distinguishable by the presence of taeniae coli, haustrations and appendices epiploicae!

Ascending colon is supplied by right colic aa from the superior mesenteric a; the middle colon is supplied by middle colic aa from the same source. But what is the source of left colic aa which supply the descending colon?

Veins usually accompany arteries in neurovascular bundles and elsewhere. The veins may be single (vena comitans) or paired (venae comitantes) and enclosed with the artery in a common sheath of connective tissue

Commissar (an official enjoined or entrusted with a particular responsibility). The anterior commisure is a narrow but distinct bundle of commissural fibres running anterior to the columns of the fornix and below the lentiform nucleus. Its fibres fan out into the anterior part of the temporal lobe, including the parahippocampal gyrus

Whilst all nerve fibres connect nerve cells to target tissues, not all are commissural in the strict sense since some are projection and others association fibres. Commissural implies connecting similar areas of left and right sides; projection implies connecting different CNS areas and association implies connecting different areas on the same side

Anatomical Names	Latin/Greek or Other Origins	Examples
Communicans, communicantes	**Communicare**: [L] *to share, communicate*	Ramus communicans
Concha, conchae	**Concha**: [L] *sea shell*	Concha(e)
Condyle, condylar	**Kondylos**: [Gr] *knuckle*	Occipital condyle
		Bicondylar joints
Conjunctiva, conjunctival,	**Conjugere**: [L] *to join together*	Conjunctiva
		Conjoint tendon
Contralateral	**Contra-**: [L] *opposite*; **latus**: [L] *side*	Contralateral limb
Coracoid, coraco-	**Korax**: [Gr] *raven*; **eidos** [Gr] *shape, form*	Coracoid process
		Coracobrachialis m
Cordis	**Cor**: [L] *heart*	Venae cordis minimae
Cornea, corneal	**Cornu**: [L] *horn, hoof, beak, claw*	Cornea

Notes, Links and Non-Anatomical Usages

Communication. The rami communicantes are nerve branches connecting spinal nn to sympathetic ganglia. Ventral rami of thoracic and upper lumbar spinal nn send a white ramus communicans to the corresponding ganglion. Ventral rami also receive a grey ramus from the corresponding ganglion. Which ramus communicans is postganglionic and what is its distribution?

Conch. The superior, middle and inferior nasal conchae are scroll-like projections from the lateral wall of the nasal cavity. They are sometimes referred to as the turbinates

Condyloid (condyle-like). A condyle is a convex prominence at the end of a bone. The knuckles are formed by the bones of the metacarpo- and inter-phalangeal joints and their convex articular surfaces. Occipital condyles articulate with the atlas at the atlanto-occipital joints. What movements occur at these joints?

These have movement mainly in one plane but with limited rotation in a plane at right angles to the first

Originally the membrana conjunctiva (conjunctive membrane), this is a mucous membrane covering the cornea at the front of the eye and lining the deep aspect of the eyelids. In this sense, it joins the front of the eyeball with the eyelids. Conjugate, conjunction, conjunctivitis, conjoint

At a number of sites, the tendons of different muscles may unite and run to a common attachment. Can you recall any?

On the opposite side of the body to the other limb

An anterolateral projection from the scapula named for its resemblance to the beak of a raven

Attaches to the coracoid process and medial humeral shaft

Cordial (warm-hearted or friendly). The venae cordis minimae are veins that drain from the myocardium directly into the atria and ventricles, mainly those on the right. The more important routes of cardiac drainage are via the coronary sinus and anterior cardiac vv

Originally the cornea tela (horny tissue), perhaps named from its continuity with the sclera of the eyeball at the corneoscleral junction

Anatomical Names	Latin/Greek or Other Origins	Examples
Cornu, cornua	**Cornu**: [L] *horn, hoof, beak, claw*	Cornua (of hyoid, thyroid, ventricles of brain)
Coronary, coronoid, coronal	**Corona**: [L] *crown, garland*	Coronary aa
		Coronoid process
		Coronal suture
Corpus, corpora	**Corpus**: [L] *body, flesh, corpse, substance*	Corpora quadrigemina
		Corpus callosum
		Corpus luteum
Cortex, cortical, cortico-	**Cortex**: [L] *rind, shell, crust, bark, cork*	Cerebral cortex
		Adrenal cortex
Costal, costo-	**Costa**: [L] *rib*	Costal cartilage
		Costoclavicular lig
Cranium, cranial, cranio-	**Cranium**: [L] *skull*	Cranium

Notes, Links and Non-Anatomical Usages

The root implies anything horn-like. Several words come from this root including cornucopia (the mythical horn of plenty), cornea (the anterior covering of the eyeball), corniculate (minor cartilages of the larynx). The hyoid and thyroid cartilages have multiple horns (cornua)

Coronation (a crowning ceremony), coronet. The coronary aa pass round the heart like a crown. The left and right coronary aa arise where? Do they anastomose?

There are two such processes: one crowns the ramus of the mandible and provides attachment for the temporalis m; the other crowns the trochlear notch on the proximal end of the ulna

Coronal planes relate to the crown of the head and divide the body vertically into anterior and posterior parts, i.e. parallel to the coronal suture of the skull which separates the frontal and parietal bones (see frontal below)

Corporation, corps, corpse, corpulent, corpuscle. The four bodies which make up the corpora quadrigemina are the two superior and two inferior colliculi. Where are they found?

The corpus callosum (tough body) is a large mass of commissural tissue joining the two cerebral hemispheres

Corpus luteum (yellow body) as opposed to corpus albicans (white body). Both bodies are found in which organ?

Cerebral cortex covers the cerebrum

Covers the adrenal gland and produces corticosteroids

Costard (a British variety of cooking apple which is large and ribbed). Costal cartilages of ribs 1-7 articulate with the sternum. With what do the cartilages of the other ribs articulate?

Runs from the inferomedial surface of the clavicle to the superior surface of the first rib and its costal cartilage. The ligament is an accessory ligament of the sternoclavicular joint

Craniates (skulled animals), craniotomy (cutting into the skull)

53

Anatomical Names	Latin/Greek or Other Origins	Examples
		Cranial nn (I to XII)
Cremaster	**Kremastos**: [Gr] *suspended, hanging*	Cremaster m
Cribriform	**Cribriform**: [L] *like a sieve or colander*	Cribriform plate
		Cribriform fascia of the thigh
Cricoid, crico-	**Krikos**: [Gr] *ring*; **eidos**: [Gr] *shape*	Cricoid cartilage
		Cricoarytenoid mm
Crista, cristae, cristal	**Crista**: [L] *a plume on crest of a helmet, crest, cock's comb*	Crista galli
		Crista terminalis
Crus, crura, crural	**Crus**: [L] *leg, shin*	Crus cerebri
		Crus of the diaphragm

Notes, Links and Non-Anatomical Usages

Cranial nn: what do O,O,O,T,T,A,F,V,G,V,A and H stand for?

The muscle raises the testis and scrotum for protection and temperature regulation and so, in a sense, suspends the testis. The nerve supply is the genitofemoral n which also supplies skin on the medial aspect of the thigh. Stroking this skin elicits the cremasteric reflex – the ipsilateral testis rises. A really neat party trick! But the reflex is useful for detecting testicular problems. Butterfly pupae have a hook-shaped structure for purposes of suspension. Guess what, it's called the cremaster!

The cribriform plate of the ethmoid bone is pierced by many little holes which transmit nerve fibres from the nasal olfactory mucosa to the olfactory bulb

Femoral superficial fascia over the saphenous opening is pierced by the great saphenous v and other vessels. This area is called the cribriform fascia

The cricoid cartilage is like a signet ring

Laterals adduct, posteriors abduct the vocal cords. Of course, you remember their nerve supply!

Crest, crestfallen. The crista galli (literally, the cock's comb) is a superior projection from the ethmoid bone in the anterior cranial fossa to which the falx cerebri attaches

Means boundary crest. It is the ridge on the interior of the right atrium running from the superior to the IVC opening. On one side of the ridge, the atrial wall is smooth; on the other, it is rough. The latter is due to muscular ridges (musculi pectinati) running anteriorly and into the auricular appendage

The leg of the cerebrum on each side forms the anterior part of a cerebral peduncle. The crura convey corticospinal projection fibres that descend through the pons and into the medulla oblongata

The leg of the diaphragm on each side attaches the diaphragm to the posterior abdominal wall. The left crus arises from L1 and L2 vertebrae. The longer right crus extends lower down (L1-L3 vertebra) and forms a sling around the lower oesophagus. The two crura unite to form a median arcuate lig behind which is the aortic hiatus (level T12, L1)

Anatomical Names	Latin/Greek or Other Origins	Examples
Crux, cruciate, cruciform	**Crux**: [L] *cross*	Posterior cruciate lig
		Cruciform lig
Cubital	**Cubitum**: [L] *elbow*	Cubital fossa
		Median cubital v
Cuboid, cuboidal	**Kubos**: [Gr] *cube*; **eidos**: [Gr] *shape*	Cuboid bone
Cuneus, cuneate, cuneiform	**Cuneus**: [L] *wedge*	Cuneiform bones
		Cuneus
		Cuneate fasciculus and tubercle
Cupola	**Cupula**: [L] *vault, small cask*	Cupola of the lung
Cusp, cuspid	**Cuspis**: [L] *point, spear, javelin*	Bicuspid (mitral) valve

Notes, Links and Non-Anatomical Usages

Crucial, crucifix, crucify, crusade (the crusaders wore tunics emblazoned with the holy cross). The anterior and posterior cruciate ligs of the knee joint appear X-shaped when viewed from the medial or lateral sides. The ligaments stabilise the joint anteroposteriorly

This has transverse and longitudinal components which hold the dens of the axis next to the anterior arch of the atlas (transverse element) and attach the atlas to the occipital bone and axis (longitudinal element)

Cubit (old measure based on forearm length), decubitus position (lying down), decubitus ulcer (bedsore). The boundaries of the cubital fossa are the pronator teres m (medial) and brachioradialis m (lateral). Between is the tendon of biceps brachii m with the brachial a on its medial side

This subcutaneous vein is so constant that it is a standard site for intravenous injection. It connects the basilic and cephalic vv of the upper limb

Cubic, Cubism (an art form pioneered by Picasso and which made use of simple geometric shapes rather than perspective). The cuboid tarsal bone is rather cuboidal in shape

Cuneiform (old form of writing using wedge-shaped symbols). When the distal articular surfaces of the three cuneiform bones of the tarsus are viewed, they look wedge-shaped. The bones, and their shape, contribute to the transverse arch of the foot

This is a wedge-shaped region on the posteromedial aspect of the occipital lobe of the cerebral hemisphere. It lies between the calcarine and parieto-occipital sulci

The cuneate tubercle is a lateral swelling at the rostral end of the cuneate fasciculus on the posterior of the medulla oblongata. The fasciculus is an ascending spinal tract and, on transverse section, looks wedge-shaped

In architecture, a cupola is a dome of a roof or ceiling. The lung and its visceral pleura have mediastinal, diaphragmatic, costal and cervical surfaces. The latter project 2-3 cm above the medial third of the clavicle as the apex, dome or cupola. Not to be confused with **copula**: [L] *union, connection, bond* (as in copulation)

The bicuspid (two cusps) is the left atrioventricular valve

Glossary 2

Anatomical Names	Latin/Greek or Other Origins	Examples
		Tricuspid valve
Cutaneous	**Cutis**: [L] *skin*	Lateral cutaneous n of forearm
Cyst, cystic	**Kuste**: [Gr] *cyst, bladder*	Cystic duct

D

Decidua, deciduous	**Deciduus**: [L] *that fallen down or off*	Decidua
Decussation	**Decussis**: [L] *the number 10 in Latin is X (decem)* **decussare**: [L] *to divide cross-wise*	Pyramidal decussation
Deep	**Deop**: [OE] *deep*	Deep
Deferens	**Defero**: [L] *carrying to a place, bringing down/ from*	Ductus deferens
Deltoid, delto-	**Delta**: [Gr] *the Greek letter Δ;* **eidos**: [Gr] *shape*	Deltoid muscle, ligament
		Deltopectoral triangle
Dens, dental, dentate	**Dens**: [L] *tooth or anything of similar shape*	Dens

58

Notes, Links and Non-Anatomical Usages

This is the right atrioventricular valve and has three cusps

A sensory n arising from the musculocutaneous n and supplying lateral forearm skin. Cuticle

Cyst, cystitis (bladder inflammation), cholecystitis (gall-bladder inflammation), cystoscope (for viewing the bladder). The cystic duct joins the gall bladder to the common hepatic duct to form the bile duct. The cystic duct has a mucous coat arranged as a spiral fold

Deciduous trees (shed their leaves), deciduous teeth (from the primary primary dentition) are shed to be replaced by secondary or permanent teeth. The uterine lining during pregnancy (the decidua) is also shed - at birth

December (the tenth month in the Roman calendar!), decimal, decimate (to reduce by a factor of 10). The pyramids are promontories on the anterior surface of the medulla oblongata. They are composed of projection fibre bundles which originate in the precentral gyrus of the cerebrum and the majority of fibres decussate (cross over to the contralateral side)

Although a familiar word, it has a very particular meaning in Anatomy which should not be forgotten: it always refers to structures lying further away from the surface of the body. Opposite to superficial

Defer, deferent[ial]. The ductus deferens carries seminal fluid from the testis

Originally, the letter symbolised the triangular opening into a tent! The delta of a river usually has this Δ shape as, not surprisingly, do the muscle and ligament

This is a small triangular depression inferior to the lateral third of the clavicle and having the deltoid and pectoralis major mm as its boundaries. It is also known as the infraclavicular fossa

Dentine, dentist, dentition, denture. The dens is the odontoid (tooth-shaped) process. Although attached to the axis, it actually represents the body of the atlas

Anatomical Names	Latin/Greek or Other Origins	Examples
		Dentate gyrus
Dermatome	**Derma + tomos**: [Gr] *slice, cut, a part cut out*	Dermatome
Dermis, dermal	**Derma**: [Gr] *skin, pelt, leather*	Dermis
Detrusor	**Detrudere**: [L] *to thrust away, evict*	Detrusor m
Diaphragm, diaphragmatic	**Dia-**: [Gr] *across*; **phragma**: [Gr] *fence*	Diaphragm
		Costodiaphragmatic recess
Digiti, digitorum, digitation	**Digitus**: [L] *finger, toe*	Flexor digiti minimi m
		Flexor digitorum longus m
Distal	**Distare**: [L] *to be distant, apart, different*	Distal
Diverticulum	**Diverticulum**: [L] *by-road, diversion, digression*	Diverticulum ilei
Dorsum, dorsal, dorsi	**Dorsum**: [L] *back*	Dorsum

Notes, Links and Non-Anatomical Usages

Part of the hippocampal complex

Dermatome (a strip of skin cut out to be supplied by nerves from a particular spinal level), myotome (a mass of skeletal muscle supplied by a particular spinal level), microtome (an instrument for cutting small slices)

Dermatology, dermatitis, pachyderms (a group of mammals with thick skin, including elephants, rhinos and…politicians?). Dermis is now just part of the skin, the other part being the epidermis (on the dermis). By the way, did you know that skin is the largest organ of the body?

The bladder wall is composed mainly of the detrusor m which evicts urine. Its motor supply is from pelvic splanchnic nn

The diaphragm separates the thoracic and abdominal cavities. Don't forget that there are also pelvic and urogenital diaphragms! What are all these diaphragms made of?

An area of pleural cavity between costal and diaphragmatic pleurae into which the lung expands during inspiration

Digital, digitigrade (toe-walker, a group of animals which, like horses, walk on their toes), prestidigitation (sleight of hand with nimble fingers!), interdigitating (interlocking like the fingers of two hands). The flexor digiti minimi m (flexor of the little finger) is an intrinsic muscle of the hand supplied by the deep branch of the ulnar n. What is the name of the eminence under which this and other little finger mm lie?

The long flexor of the four lateral toes. It runs from the tibia distal to the soleal line and runs to the distal phalanx of a toe. It also helps to plantar flex at the ankle joint. Nerve supply?

Distance, distant. The distal end of a limb is the part furthest from the limb root

Diversion, diverticulitis. The diverticulum ilei (or Meckel's diverticulum) is an embryological remnant which may contain ectopic gastric tissue

Dorsiflexion. Dorsal and posterior are used interchangeably

Anatomical Names	Latin/Greek or Other Origins	Examples
		Latissimus dorsi m
		Dorsal root
Duodenum, duodenal	**Duodeni**: [L] *numbered or divided in units of 12*	Duodenum
Dura, dural	**Durus**: [L] *hard, stiff, lasting*	Dura mater
		Dural reflections
		Epidural

E

Anatomical Names	Latin/Greek or Other Origins	Examples
Efferent	**Efferens**: [L] *carrying away*	Efferent nerve Efferent lymphatic
Epididymis	**Epi-**: [Gr] *near, upon*; **didymos**: [Gr] *twin, testicle*	Epididymis
Epiglottis, epiglottic	**Epi-**: [Gr] *near, upon*; **glossa/glotta**: [Gr] *tongue*	Epiglottis
		Aryepiglottic folds

Notes, Links and Non-Anatomical Usages

The very broad muscle of the back is innervated by the thoracodorsal n and is a medial rotator, extensor and adductor of the arm at the shoulder joint. Where does it insert on the humerus?

Contains afferent nerve fibres passing information to the spinal cord. It bears a swelling, the spinal or dorsal root ganglion, which contains unipolar neurons with bifurcating dendro-axonal processes

Duodenum (this is traditionally 12 finger-widths long), duodecimal

Durable (hard-wearing, lasting), endurance (ability to last), duration (the period over which something endures), Durex (tradename for a presumably hard-wearing condom), obdurate (refusing to change an opinion or action, hard-hearted). The dura mater is the tough, outermost layer of the three meninges covering brain and spinal cord

A term used to describe the arrangement of the dura mater within the cranial cavity

Literally, on or around the dura mater. Such an injection is made into the space surrounding the dura mater, usually of the spinal cord

Efferent – carrying away from (e.g. the CNS, an organ or a lymph node)

Epididymis (near the testicle) – a firm mass of coiled tubes about 6 metres long lying posterior to the testis. It has a head (caput), body (corpus) and tail (cauda). If there's anyone reading this by the name of Thomas, your name comes from an Aramaic word-root for twin. And that's not a load of ....!

It is near the tongue. What is its function?

Run from arytenoid to epiglottic cartilages

Anatomical Names	Latin/Greek or Other Origins	Examples
Epiploic	**Epiploon**: [Gr] *bowels, membrane*	Epiploic foramen
Ethmoid, ethmoidal	**Ethmos**: [Gr] *sieve*; **eidos**: [Gr] *shape, form*	Ethmoid bone
		Sphenoethmoidal recess
Eversion, evertor	**E[x]-**: [L] *out*; **vertere**: [L] *to turn*	Eversion
		Evertor mm
Extension, extensor	**Extendere**: [L] *extend, straighten, stretch out*	Extension Extensor muscles
Externus	**Externus**: [L] *external, outward, foreign, strange*	Obliquus externus m

F

Falx, falciform	**Falx**: [L] *sickle, scythe*	Falx cerebri
		Falciform lig

Notes, Links and Non-Anatomical Usages

Epiploon is equivalent to omentum (see below). The epiploic foramen is the communication between the lesser and greater peritoneal sacs. Its anterior wall is the free edge of the lesser omentum which contains three important vessels: bile duct on the right, hepatic a on the left and portal v behind and between

So... the cribriform plate of the ethmoid bone is the sieve-like plate of the sieve bone!

A small region of the nasal cavity above the superior concha and into which opens the sphenoidal and, maybe, posterior ethmoidal air sinuses

Eversion of the foot involves turning the sole of the foot outwards (laterally). At what joints does this occur?

The set of muscles which produces eversion of the foot, i.e. peroneus longus m and ...?

Extend, extent, extensive, extenuate. Extension usually involves straightening out the joint and is brought about by the action of extensor mm

The external oblique mm of the abdomen have fibres running in the same direction as the fingers when the hands are in trouser pockets, i.e. downwards and medially. The aponeurosis helps to form which important ligament running between the anterior superior iliac spine and the pubic tubercle?

Plasmodium falciparum (the infectious agent of the most severe form of malaria is sickle-bearing), falcon (because the beak is sickle-shaped?). The falx cerebri is the sickle-shaped fold of dura mater which contains the sagittal dural venous sinuses and helps limit side-to-side and antero-posterior movement of the brain within the cranial cavity

The sickle-shaped ligament is a peritoneal fold running from the umbilicus up to the portal area of the liver. In its free border is the ligamentum teres, the remains of the umbilical v which delivered oxygenated blood to the fetus in utero

Anatomical Names	Latin/Greek or Other Origins	Examples
Fascia, fasciae, fascial	**Fascia**: [L] *band, bandage*	Superficial/deep fascia
Fascicle, fasciculus, fasciculi, fascicular	**Fasciculus**: [L] *little bundle*	Fasciculus
Fauces	**Fauces**: [L] *throat, jaws, pass, straits, isthmus*	Fauces
Femur, femoral, femoris	**Femur**: [L] *thigh*	Femur
		Femoral artery
		Biceps femoris m
Fenestra, fenestrae	**Fenestra**: [L] *window, wall opening*	Fenestra cochleae Fenestra vestibule
Fetus, fetal	**Fetus**: [L] *bringing forth fruit or young*	Fetus
Fibula, fibulae, fibular[is]	**Fibula**: [L] *clasp, buckle, brooch, pin*	Fibula
		Fibular

Notes, Links and Non-Anatomical Usages

Fascia or facia. Superficial fascia is synonymous with the histological term, hypodermis. In the abdominal wall, it is divisible into a superficial fatty layer and a deep membranous layer. The latter forms the superficial layer of the superficial perineal pouch

The gracile and cuneate fasciculi are ascending tracts of nerve fibres in the posterior part of the medulla oblongata

Fauces (area between oral cavity and pharynx). The fauces are bounded on each side by mucosal ridges, the palatoglossal folds. Between these and the more posterior palatopharyngeal folds, the palatine tonsils normally lie

Femur (bone of thigh), femoral (relating to the femur or thigh)

The femoral a enters the thigh behind the inguinal lig at a point midway between the anterior superior iliac spine and the pubic symphysis. The pulse of the artery can be palpated at this site by pressing gently backwards against pectineus m and the superior pubic ramus

'The two-headed thigh muscle'. So-called to distinguish it from biceps brachii m

Fenestrated (such capillaries have openings in their endothelium), defenestration (throwing someone out of a window!). The fenestrae cochleae and vestibuli are two windows in the medial wall of the tympanic cavity. The former is closed by a membrane and looks on the scala tympani of the cochlea. The fenestra vestibuli leads into the vestibule of the inner ear and the base of the stapes fits snugly into it so that vibrations of the stapes may be transmitted to the inner ear

Not to be confused with **foetus**: [L] *stinking*! So the correct spellings are fetus/fetal and not foetus/foetal

Fibula (a pin – if you look at the tibia and fibula together, especially in a small mammal such as a rabbit, they look like a clasp with the tibia being the catch and the fibula the pin)

Fibular is sometimes used to mean on the lateral side (of the leg or ankle). The peroneal mm are sometimes called the fibular mm

Anatomical Names	Latin/Greek or Other Origins	Examples
Filum, fila, filiform	**Filum**: [L] *thread, string, fibre*	Filum terminale
		Filiform papillae
Fimbria, fimbriae	**Fimbria**: [L] *fringe, border*	Fimbria of the uterine tube
Fistula, fistulae	**Fistula**: [L] *pipe, tube*	Anal fistula
Flexion, flexor	**Flexio**: [L] *bending*	Flexion Flexor muscles
Flocculus, floccular	**Floccus**: [L] *flock, lock of unspun wool*	Flocculonodular lobe
Folium, folia, foliate	**Folium**: [L] *leaf*	Folia of the cerebellum
		Foliate papillae
Follicle, follicular	**Folliculus**: [L] *little bag*	Lymphoid follicle

Notes, Links and Non-Anatomical Usages

Filament (thin wire thread in a light-bulb), fillet (a thin boneless strip of meat), filaria (a group of parasitic nematodes also known as the threadworms!). In adults, the spinal cord ends at L1-L2 vertebral level. From this point, a prolongation of the pia mater, the filum terminale (the terminal thread), descends to attach to the coccyx

Filiform (thread-like) papillae are found on the dorsum of the tongue and help to increase frictional contact with food

Fimbriated (fringed). The fimbriae of the uterine tube arise from the infundibulum. One fimbria, the ovarian fimbria, is longer than the others and is applied to the tubal pole of the ovary

A fistula is an abnormal or surgically-made passage between one organ and either another organ or the skin surface. Infection of the mucosa of the anal canal can spread into the ischiorectal (or ischio-anal) fossae from where it can open onto the skin surface

Flex, flexible. Flexion usually involves bending at a joint. It is effected by flexor mm. In the case of the ankle, there is plantar flexion and dorsiflexion. Plantar flexion is assisted by long toe flexors but dorsiflexion is assisted by toe extensors

Flocculent (aggregated in woolly, cloudlike masses), a flock mattress is stuffed with wool waste or torn-up cloth, flock wallpaper bears powdered wool or cloth. The flocculonodular node of the cerebellum is on its inferior aspect and is composed of two lateral tufted outgrowths, the flocculi, and a single midline nodule. Together they are concerned with maintaining balance (equilibration)

Foliage, folic acid (found in green leaves), portfolio (a case for carrying important papers). The cerebellar surface is divided into folds (folia; singular folium) and grooves (fissures) which are the counterparts of cerebral gyri and sulci respectively

The mucosa of the superior surface of the tongue bears taste, touch and pressure receptors. Foliate papillae are reddish, leaf-like and tend to lie at the sides at the junction of its oral and pharyngeal parts. They contain taste buds

Lymph node cortex contains dense aggregates of lymphocytes (lymphoid follicles). Follicles may be primary or secondary and the lymphocytes are mainly B with some helper T cells

Anatomical Names	Latin/Greek or Other Origins	Examples
		Hair follicle
Foramen, foramina	**Foramen**: [L] *hole, opening, aperture*	Foramen magnum
		Foramen lacerum
		Foramen ovale
		Foramen rotundum
		Foramen spinosum
Fornix, fornices	**Fornix**: [L] *arch, vault, brothel situated therein*	Fornix
Fossa, fossae	**Fossa**: [L] *ditch, trench, recess*	Fossa ovalis
		Cranial fossae
Fovea	**Fovea**: [L] *pit, snare*	Fovea centralis
		Pterygoid fovea
Frenulum	**Frenum**: [L] *bridle, bit, band*	Frenulum (of tongue, lips, prepuce, clitoris)

Notes, Links and Non-Anatomical Usages

The epidermal pit through which the hair shaft grows to reach the surface of the skin

Foraminifera (protozoa with shells having numerous openings in them). The foramen magnum is just that – a big hole

The mangled or torn hole, transmitting what?

The oval hole, transmitting what?

The rounded hole, transmitting what?

The spine-related hole because of its nearness to the spine of the sphenoid bone. What does it transmit?

Fornication (Roman prostitutes entertained clients under the arches of the Colosseum and other buildings!). The fornices of the vagina encircle the vaginal cervix and provide important sites via which surrounding pelvic structures may be palpated

Fossil (often dug up), fossorial (having a digging habit, like a mole). The fossa ovalis is a depression in the inferior part of the interatrial septum of the heart. It is the remnant of the foramen ovale which shunted blood away from the pulmonary circulation in the fetal heart

There are anterior, middle and posterior cranial fossae. Which brain regions are found in each?

On the posterior of the retina is an oval macula lutea (yellow spot) which is the area of greatest visual acuity. It has a central depression, the fovea centralis (central pit)

A triangular depression on the anterior of the neck of the mandible into which the lateral pterygoid m is inserted. What are the other attachments of this muscle?

In all cases, the frenulum forms a fold which checks the range of movement of the structure to which it attaches

Anatomical Names	Latin/Greek or Other Origins	Examples
Frontal, frontalis	**Frons**: [L] *front, forehead*	Frontal bone
		Occipitofrontalis m
Fundus	**Fundus**: [L] *foundation, bottom*	Fundus (of bladder, stomach, uterus)
Fungiform	**Fungus**: [L] *fungus, mushroom*	Fungiform papillae
Funiculus, funiculi	**Funiculus**: [L] *thin or slender rope*	Spinal funiculi
G		
Galea	**Galea**: [L] *helmet*	Galea aponeurotica
Ganglion, ganglionic	**Ganglion**: [Gr] *tumour, swelling*	Spinal ganglion
		Ganglion impar
		Pre- or post-ganglionic
Gastric, gastro-	**Gastrikos**: [Gr] *related to the stomach*	Gastric vv

Notes, Links and Non-Anatomical Usages

Front, frontispiece. A frontal plane passes vertically like a coronal plane (see above). The frontal bone is the bone of the forehead. With which other skull bones does it articulate?

A muscle of facial expression with occipital and frontal bellies joined by a tough aponeurosis of the scalp (the epicranial aponeurosis). Contraction of the frontal bellies on both sides raises the eyebrows

Profound (deep), the fundus is the base of an organ when viewed from the traditional surgical approach. Foundation, fund, fundament (also a humorous term for the buttocks or bottom!)

Large, round, deep-red taste buds found on the dorsum of the tongue but mainly at its lateral margins

Funicular (a cable railway). The term is used to describe cord-like structures or nerve bundles. The spinal funiculi are three bands of white matter – the anterior funiculus, lateral funiculus and posterior funiculus. What do they convey? The term funiculus has been applied also to the umbilical cord and spermatic cord

The epicranial aponeurosis (= galea aponeurotica) is a flat tendon joining the frontal and occipital bellies of the occipitofrontalis m. It constitutes the third layer of the scalp. Memorise this acronym: SCALP = Skin, Connective tissue, Aponeurosis, Loose connective tissue, Pericranium

Ganglion – a swelling caused by a collection of neurons. The spinal ganglion is a sensory ganglion associated with the dorsal spinal root

Impar: [L] *different, unequal.* The ganglion impar is formed anterior to the coccyx by the meeting of the two sympathetic trunks. It is single and unpaired and, therefore, different from other ganglia

Terms used to distinguish fibres running to or from a ganglion

Gastrointestinal, gastronomic, gastrula. Left and right gastric vv drain the stomach and lesser omentum and empty into the portal v. Where do the short gastric vv drain and empty?

Anatomical Names	Latin/Greek or Other Origins	Examples
		Epigastric region
		Gastrocnemius m
		Digastricus m
Gemellus, gemelli	**Gemellus**: [L] *little twin*; **geminus**: [L] *twin*	Gemelli mm
Genicular, geniculate	**Geniculatus**: [L] *jointed, having knots* **geniculum**: [L] *little knee*	Genicular branches of the obturator n
		Geniculate ganglion of facial n
Genioid, genio-	**Geneion**: [Gr] *chin*	Genioid tubercles
		Genioglossus m
Genu	**Genu**: [L] *knee*	Genu of corpus callosum
Gingiva, gingivae, gingival	**Gingiva**: [L] *gum*	Gingiva
Glabella	**Glaber**: [L] *bald, smooth*	Glabella

Notes, Links and Non-Anatomical Usages

The epigastric region or epigastrium (near the stomach) is the upper middle region of the anterior abdominal wall. The lower middle region is the hypogastrium (below the stomach)

Gastrocnemius m (belly-leg, i.e. calf, muscle). What are its attachments?

Has two bellies!

The gemelli twins are the superior and inferior gemellus mm. These very small muscles are accessory lateral rotators of the hip joint. The constellation Gemini contains the heavenly twins, Castor and Pollux

Genicular branches of the obturator n pass to the knee joint (together with other branches from which named nerves?)

This ganglion is found on the site (called the geniculum) where the facial (cranial VII) n alters its course. The ganglion is sensory and associated with which special sense?

Genial (related to the chin). The genioid tubercles lie on the posterior aspect of the mental area of the mandible and are sometimes called the mental (**mentum**: [L] *chin*) spines or tubercles. They are attachment sites for important muscles. Here's one now!

Pulls the tongue forwards and out of the mouth. This is called protrusion of the tongue

Genuflect (to pray or go down on bended knee), genuine (when a Roman child was born, it was the practice for the father to claim it as his own by putting it on his knee!). The genu of the corpus callosum is at its anterior pole where, on a sagittal section, it looks as though the white matter bends. The posterior pole of the corpus callosum is called what?

Gingivitis (inflammation of the gums)

Glabrous (free from hair, smooth). The glabella is the slight midline elevation between the eyebrow ridges and superior to the depressions at the root of the nose. It is usually (but not always) free of hair

Anatomical Names	Latin/Greek or Other Origins	Examples
Gland, glans	**Glans**: [L] *acorn, nut, bullet;* **glandula**: [L] *a term for swollen glands in the neck*	Glans penis
		Salivary glands
Glenoid, glenoidal, gleno-	**Glene**: [Gr] *socket;* **eidos**: [Gr] *shape, form*	Glenoid fossa/cavity
		Glenoidal labrum
		Glenohumeral ligs
Glossal, glosso-, -glossus, glottis, glottidis	**Glossa** or **glotta**: [Gr] *tongue*	Hypoglossal n (cranial XII)
		Glossopharyngeal n (cranial IX)
		Hyoglossus m
		Rima glottidis
		Epiglottis
Gluteus, gluteal	**Gloutos**: [Gr] *buttock, rump*	Gluteus maximus m
		Gluteal lines
Gracilis	**Gracilis**: [L] s*lender, lean*	Gracilis m
Gubernaculum	**Gubernaculum**: [L] *helm, rudder, tiller* **gubernator**: [L] *governor*	Gubernaculum testis

Notes, Links and Non-Anatomical Usages

The glans penis is the swelling at the distal end of the penis. Its skin is removed during circumcision ("The unkindest cut of all?")

The paired parotid, submandibular and sublingual salivary glands

The glenoid fossa is an articular socket for the humeral head

The labrum is the fibrocartilaginous rim of the glenoid fossa

These are weak ligaments reinforcing the joint capsule anteriorly

Glossary, glottal. The hypoglossal n supplies all intrinsic and most extrinsic mm of the tongue. What are the exceptions?

Leaves the cranial cavity through which foramen? It supplies taste to the posterior two-thirds of the tongue and contributes to the pharyngeal plexus. It innervates only one skeletal muscle. Which one?

Runs from hyoid bone to tongue and depresses the latter

The rima glottidis is the gap between the vocal cords

Lies near the tongue (but see rima below)

Gluteus maximus is the largest of three gluteal mm. The others are?

Again, there are three and they are associated with the origins of gluteal mm. Can you recall them?

Gracile (= slender). *Homo gracilis* was of slender build. The gracilis m is a long, slender muscle which is part of the adductor group of the thigh. So what is its innervation?

The gubernaculum testis 'steers' the testis in its descent from the posterior abdominal wall, via the inguinal canal, and into the scrotum. In the USA, gubernatorial elections decide who the state governors will be

77

Anatomical Names	Latin/Greek or Other Origins	Examples
Gyrus, gyri	**Gyrus**: [L] *circle, coil, ring*	Precentral gyrus
		Gyrencephalic

H

Hamate	**Hamus**: [L] *hook*	Hamate bone
Hamulus	**Hamulus**: [L] *little hook*	Pterygoid hamulus
Haustra, haustration	**Haustrum**: [L] *bucket, pouch*	Haustrations
Hiatus	**Hiatus**: [L] *gap, opening*	Adductor hiatus
		Hiatus semilunaris
Hilum, hilus, hilar	**Hilum**: [L] *the scar on a seed where it joins its stalk*	Hilum (of liver, lung, spleen, etc)
Hippocampus, hippocampal	**Hippos**: [Gr] *horse;* **kampos**: [Gr] *sea creature*	Hippocampus
Humour	**Humor**: [L] *liquid, moisture*	Aqueous humour and vitreous humour

Notes, Links and Non-Anatomical Usages

Gyrate (spin round), gyroscope. The precentral gyrus of the cerebrum is the motor area and lies anterior to the central sulcus. What gyrus lies immediately behind the sulcus?

The human brain is gyrencephalic (much folded by gyri and sulci) in contrast to that of smaller mammals (e.g. mouse) which is lissencephalic (relatively smooth and unfolded)

The hamate bone is a carpal bone with a hook to which the flexor retinaculum attaches

Hamulus – used to describe hook-like appendages. The pterygoid hamulus is a small hook-shaped piece of bone hanging from the inferior of the medial pterygoid plate. You should be able to palpate this at the posterolateral angle of your own hard palate!

Amongst the distinguishing features of the large intestine are the taeniae coli, the appendices epiploicae and the sacculations or haustrations. The latter give the colon a pouch-like or segmented appearance. Loss of haustra is seen in ulcerative colitis

Hiatus hernia is herniation of part of the stomach through an anatomical gap in the diaphragm and into the thoracic cavity. The adductor hiatus is a gap in the adductor magnus tendon at which the femoral a becomes the popliteal a. Does anything else pass through this hiatus?

A groove on the medial wall of the nasal cavity

The hilum (or hilus) of an organ is the site at which its vessels (mainly blood vessels) enter and leave, so you should be able to appreciate the analogy

The hippocampus of the forebrain was named because, in a coronal section, it resembles a seahorse (generic name *Hippocampus*)

Humid, humorous. In less enlightened days, the balance of the humours (the four bodily fluids – blood, phlegm, yellow bile and black bile) was supposed to affect one's mental state or mood. I like to think that a good laugh, prompted by a funny joke, is humorous because it makes your eyes water!

79

Anatomical Names	Latin/Greek or Other Origins	Examples
Hymen	**Hymen**: [Gr, L] *God of Marriage*, [Gr] *membrane*	Hymen
Hyoid, hyo-	**Hyo-**: [Gr] *related to U, the letter upsilon*; **eidos**: [Gr] *shape, form*	Hyoid bone
		Hyoglossus m
Hypophysis, hypophyseal	**Hypo-**: [Gr] *below, down*; **physis**: [Gr] *growth*	Hypophysis
Hypothenar	See Thenar below	
I		
Ileum, ileal, ilei, ileo-	**Ile**: [L] *gut*; **ilia**: [L] *entrails, loins*	Ileum
		Ileal aa
		Ileocolic aa
		Diverticulum ilei
Ilium, iliac, ilio-	**Ilia**: [L] *entrails, loins*	Ilium (iliac bone)
		Iliolumbar lig
Ima	**Ima**: supposedly from [L]: *lowest*	Thyroidea ima a

Notes, Links and Non-Anatomical Usages

Hymeneal (relating to marriage), hymenoptera (insects, e.g. bees and wasps, with two pairs of membranous wings). The hymen is a thin mucous membrane located at the vaginal orifice. Its rupture during the first experience of sexual intercourse leaves remnants around the vaginal opening

The hyoid bone is U-shaped with the body and two lesser horns lying anteriorly and two greater horns posteriorly

The hyoglossus m runs from the hyoid bone to the tongue. It is an extrinsic muscle of the tongue which depresses it (nerve supply?)

Diaphysis, epiphysis, metaphysis. The hypophysis cerebri (to give its full name) is a downgrowth or offshoot of the cerebrum commonly known as the pituitary gland

Ileostomy, ileitis. Here, ileum refers to entrails (bowels, guts)

Branches of the superior mesenteric a

Branches of the superior mesenteric a

Meckel's diverticulum: in 2% of subjects, 2 feet from the ileocolic junction and about 2 inches long! It may contain ectopic acid-secreting tissue and mimic symptoms of appendicitis

Here, the term ilium refers to the loins. The ilium is at the inferior end of the anatomical loins (running from 12th rib to iliac crest) but superolateral to the poetic loins (the private parts)

The iliolumbar lig runs from the transverse process of the L5 vertebra to the iliac crest. It is the inferior attachment of the quadratus lumborum m

The lowest (and smallest and least constant) of the arteries supplying the thyroid gland. It may arise from the right common carotid, subclavian or internal thoracic aa or from the aorta. What are the more important sources of arterial supply to the thyroid gland?

81

Anatomical Names	Latin/Greek or Other Origins	Examples
Incisor, incisive, incisure	**Incidere**: [L] *to cut into*	Incisor tooth
		Incisive fossa
		Angular incisure or notch
Incus	**Incus**: [L] *anvil*	Incus
Index, indicis	**Index**: [L] *fore-finger, informer, witness, spy*	Index finger
		Extensor indicis m
		Radialis indicis a
Infundibulum, infundibular	**Infundibulum**: [L] *funnel, funnel-shaped passage*	Infundibulum (of uterine tube, hypothalamus)
Inguinal, inguinalis	**Inguen**: [L] *groin*	Inguinal region
		Inguinal lig
		Falx inguinalis
Internus	**Internus**: [L] *internal, civil, domestic*	Obliquus internus m
Intimi, intima	**Intimus**: [L] *innermost, deepest*	Intercostales intimi mm

Notes, Links and Non-Anatomical Usages

Chisel-like front teeth specialised for cutting. How many are there in each jaw in adults? The same word root gives us incision, incisive and incisure

Lies posterior to the upper incisor teeth in the hard palate. Its walls contain incisive foramina which pass between the nasal cavity and hard palate. What nerves do they transmit?

A gastric notch, varying in position with stomach distension, lying on the lesser curvature. It lies between the body and pyloric antrum

Incuse (the design stamped on a coin). The incus is one of the three auditory ossicles, malleus (hammer), incus (anvil) and stapes (stirrup). All easily identifiable by their shapes!

Indicate (to point with the index finger!), indicative

An extensor of the index finger. Supplied by the radial n

The artery on the lateral (radial) side of the index finger. It arises from the radial a in the palm as it passes through the first dorsal interosseous m

Infundibular. Whether part of the uterine tube or hypothalamus, it is still funnel-shaped

The region where the thigh joins the abdomen

The inrolled inferior margin of the aponeurosis of the external oblique m which helps define the inguinal canal

Another name for the sickle-shaped conjoint tendon of the fused fibres of internal oblique m and transversus abdominis m which runs to the pubic tubercle, crest and pecten

Internal. Obliquus internus (internal oblique) mm of the abdomen have fibres running at right angles to those of external oblique mm

Intimacy, intimate. The intercostales intimi mm are sometimes called the innermost intercostal mm. Like external and internal intercostal mm, those of a given intercostal space are supplied by the corresponding intercostal n but between which pair of muscles does the nerve run?

Anatomical Names	Latin/Greek or Other Origins	Examples
		Tunica intima
Inversion, invertor	**In**-: [L] *in, inwards*; **vertere**: [L] *to turn*	Inversion
		Invertor mm
Ipsilateral	**Ipse**: [L] *self, same*; **latus**: [L] *side*	Ipsilateral
Ischium, ischial, ischio-	See Sciatic below	
Isthmus	**Isthmus**: [L from Gr] *neck, passage*	Thyroid isthmus

J

Jejunum, jejunal	**Jejunus**: [L] *empty, hungry*	Jejunum
		Jejunal vv
Jugular	**Jugulum**: [L] *neck, throat, collar* **jugum**: [L] *yoke*	Jugular notch

L

Labium, labia, labial	**Labium**: [L] *lip*	Labium major
Labyrinth, labyrinthine	**Labyrinthos**: [Gr] *maze, labyrinth*	Labyrinth

Notes, Links and Non-Anatomical Usages

The innermost layer of a blood vessel wall composed of squamous endothelial cells. What are the middle and outer layers called and what are their principal tissue types?

Inversion of the foot is when the sole turns inwards

The invertors of the foot include tibialis anterior m (nerve supply?) and … ?

Ipsilateral (on the same side) is in contrast to contralateral. Two Latin expressions still in use: ipse dixit (he himself said it – refers to an unproven or dogmatic assertion); ipso facto (by that very fact)

In ancient Greece, the Isthmian Games were held each year at Corinth which is situated on the isthmus joining central Greece to the Peloponnese peninsula. The isthmus of the thyroid gland connects the two lobes and usually lies anterior to the 2nd and 3rd tracheal cartilages

Jejune (naïve, insipid, dull). Jejunum is actually an abbreviation of intestinum jejunum (empty intestine) derived from the observation that often, after death, the jejunum is empty

Drain into the superior mesenteric vv and then into which vein?

Jugulate (to slit the throat!), subjugate (put under the yoke) Yoke itself comes from the word-root jugum. A yoke is a wooden frame attached to the neck of oxen or worn on a person's shoulders to carry things. The jugular notch is an alternative name for the suprasternal notch. Just above it, the two anterior jugular vv are connected to each other by a vein called the jugular arch

Labials (sounds produced by using the lips). The major labia are the larger of the two sets of lips associated with the female external genitalia

The labyrinth of the ear has bony and membranous parts and is a complex subserving hearing and balance

85

Anatomical Names	Latin/Greek or Other Origins	Examples
		Labyrinthine aa
Labrum	**Labrum**: [L] *lip, edge, rim*	Glenoidal labrum
Lacrimal	**Lacrima[e]**: [L] *tear[s]*	Lacrimal gland
Lacuna, lacunae, lacunar	**Lacuna**: [L] *pit, hole, chasm* **lacus**: [L] *lake, pond*	Osteocyte lacuna(e)
		Lacunar lig
Lambda, lambdoid	**Lambda**: [Gr] *Greek letter* Λ; **eidos**: [Gr] *shape*	Lambda
		Lambdoid suture
Lamina, laminae, laminar	**Lamina**: [L] *plate, leaf, blade, thin piece of metal*	Cricoid lamina Thyroid lamina
Lateral, lateralis, lata[e]	**Latus**: [L] *side, flank*	Lateral head of triceps m

Notes, Links and Non-Anatomical Usages

The artery supplies the inner ear and is usually a branch of the basilar a or the anterior inferior cerebellar a and it accompanies which cranial nn into the internal acoustic (auditory) meatus?

Labret (any body-piercing ornament attached to a lip). The glenoidal labrum is the rim of the glenoid fossa of the scapula

Lacrimation (tear flow, i.e. weeping), lacrimatory (something which causes lacrimation, e.g. a freshly-sliced onion, a good joke or a kick in the …!). The gland is situated in the superolateral part of the orbit but lacrimal fluid drains from a duct found at the medial angle of the eye. So tears run across the eye downwards and medially

Lagoon (**laguna**: [It,Sp]) is from this word-root as is the word lake. Each osteocyte (bone cell) dwells in a small cavity called a lacuna but the cells remain multiply interconnected by cell processes

The lacunar lig is at the medial end of the inguinal lig and may be so-called because of the space between the inguinal lig and the pelvic bone or because its free edge is the medial margin of the femoral ring which leads into the femoral canal

The area of the skull where the sagittal and lamboid sutures meet (so corresponding to the apex of the letter Λ). In neonates, the parietal and occipital bones are partially unfused and the posterior fontanelle is found at this site

The lambdoid suture between the occipital and parietal bones is lambda shaped

Laminated (laminated wood is made of several layers bonded together), laminin (a protein of the basal lamina). The cricoid lamina is the flat posterior plate of cartilage. The thyroid laminae on both sides meet anteriorly at an angle which is sharper in males than females. And that's why the "Adam's apple" is more prominent in males!

The lateral head of triceps m arises from the upper and lateral half of the posterior humeral shaft above the radial (or spiral) groove. A common tendon for all three heads of triceps inserts on the olecranon process of the ulna

Anatomical Names	Latin/Greek or Other Origins	Examples
		Vastus lateralis m
		Tensor fasciae latae
Latissimus	**Latus**: [L] *broad, extensive, wide, copious* **latissimus**: [L] *very broad, wide*	Latissimus dorsi m
Lentiform	**Lens**: [L] *lentil*; **forma**: [L] *shape*	Lentiform nucleus
Levator	**Elevare**: [L] *to lift, raise*	Levator scapulae m
Lieno-	**Lien**: [L] *spleen*	Lienorenal lig
Ligament, ligamentum	**Ligamen**: [L] *band, tie*	Ligamentum nuchae
Limbus, limbic	**Limbus**: [L] *border, fringe, hem, stripe, band*	Limbic lobe
		Limbic system

Notes, Links and Non-Anatomical Usages

This muscle arises from the intertrochanteric line, base of greater trochanter and linea aspera of the femur. It inserts into the tendon of quadriceps femoris m and, hence, into the patella. What is its nerve supply?

Literally, the muscle that tenses the fascia lata ('band on the flank'). This thigh muscle arises from the anterior iliac crest and anterior superior iliac spine to insert on the anterior border of the iliotibial tract of the fascia lata. It tenses the iliotibial tract, helps to extend at the knee and medially rotates at the hip. It also braces the knee, especially when the contralateral limb is off the ground. What is its nerve supply?

Defining position on the Earth requires angles of latitude and longitude. The latissimus dorsi m is the very broad muscle of the back. It has an origin from the dense thoracolumbar fascia and inserts on the humerus at the intertubercular groove. What are its actions and nerve supply?

Lens (named because of its shape resemblance to a lentil), lenticular or lentoid (biconvex, like a lentil). The lentiform nucleus is part of the corpus striatum and comprises a medial and pale globus pallidus (pale globe) and a darker lateral putamen (meaning shell or husk)

Elevate, elevator (a lift), levitation. Levator mm generally raise (e.g. levator scapulae raises the scapula) or support (levator ani supports pelvic viscera)

The lienorenal lig is a peritoneal reflection from the left kidney to the spleen. The splenic a runs within its layers

Ligament (ties bones together), ligand, ligature. The ligamentum nuchae (ligament of the nape of the neck) runs from the occiput down to the spine of C7 vertebra. It binds cervical spinous processes together and acts as an attachment site for cervical musculature

Limbo (the borders of Hell!). The limbic lobe is a collective term often used to describe the cingulate gyrus + parahippocampal gyrus + uncus. These cerebral structures border the corpus callosum and 3rd ventricle

The limbic system is a complex which includes the limbic lobe and medial parts of the frontal lobe interconnected with the hypothalamus, thalamus, basal ganglia and other deep nuclei. It is involved in emotional expression and experience

Anatomical Names	Latin/Greek or Other Origins	Examples
Linea	**Linea**: [L] *line, string, boundary*	Linea alba
		Linea aspera
Lingual	**Lingua**: [L] *tongue, language*	Lingual n
Lingula, lingulae	**Lingula**: [L] *little tongue*	Lingula (of mandible, tongue, cerebellum)
Lumbar, lumborum, lumbo-	**Lumbus**: [L] *loin*	Lumbar plexus
		Quadratus lumborum m
		Lumbosacral trunk
Lumbrical	**Lumbricus**: [L] *worm*	Lumbrical mm
Lunate, lunar, lunaris, lunule	**Luna**: [L] *moon*	Lunate bone
		Semilunar valves

Notes, Links and Non-Anatomical Usages

The white line, a tendinous raphe running from the xiphisternum to the pubic symphysis

The rough or rugged line is a broad ridge on the posterior border of the shaft of the femur. It forms an attachment for the adductor magnus m and what other muscles?

Linguistics, language (via **langue**: [Fr] *tongue*). Ironically, the phrase lingua franca (French tongue – referring to a common language used by people with different Mother tongues) is Italian and not French! The lingual n (a branch of which cranial n?) supplies taste to the anterior two-thirds of the tongue

The lingula of the mandible is a tongue-like bony projection which lies next to the mandibular foramen and to which the sphenomandibular lig is attached. The lingulae of the lung and cerebellum are also tongue-like

Lumbar puncture, lumbago (low back pain). Anatomically, the loins extend from the false ribs to the iliac crest. A loin chop includes the psoas major m (see below) and part of the adjacent lumbar vertebra. The lumbar plexus is associated with ventral primary rami of spinal nn L1-L4 supplemented by a branch of T12. What named nerves arise from this plexus?

A quadrate muscle attaching to the twelfth rib, iliolumbar lig and iliac crest

Branches of the 4th and 5th lumbar spinal nn of the lumbar plexus which descend on the pelvic sacrum to join the sacral plexus

Lumbricus terrestris is the common earthworm. The lumbrical mm of each hand (four in number) are worm-like muscles. They originate on tendons of flexor digitorum profundus mm and insert on lateral sides of metacarpophalangeal joints of digits 2-5 (so what are their actions?). Their innervations mirror the nerve supplies of flexor digitorum profundus m: the medial two lumbricals are supplied by the ulnar n and the lateral two by the median n

Lunar, lunate, lunatic (whose sanity was believed to be affected by the lunar cycle!). The lunate bone is crescent-shaped so the term does not refer just to the full moon

Here, it is a half moon shape that is being referred to

Notes, Links and Non-Anatomical Usages

A curved groove on the medial wall of the nasal cavity created, in part, by the bulge of the ethmoidal bulla

A diminutive of **luna** to mean a moon-like or crescent shape. Used to define the whitish crescent seen at the base of the finger nail (although that of the little finger is usually hidden by overlying skin). Also seen on toe nails, it represents the visible part of the nail root. Lunule is also used to describe segments of the free borders of the cusps of semilunar valves. If these thicken or become damaged by disease, the valves do not close properly and become incompetent

Immaculate (literally, spotless). The macula lutea (yellow spot) of the retina is a yellowish spot with a depression at its centre (the fovea centralis) which is the region of greatest visual acuity

The medial malleolus (a tibial prominence) is slightly superior to the lateral (a fibular prominence). Both malleoli resemble little hammer-heads. Behind the medial malleolus are several important structures and the muscle tendons can be remembered as Tom, Dick and Harry moving posteriorly (Tom is Tibialis posterior tendon, Dick is flexor Digitorum longus tendon and Harry is flexor Hallucis longus tendon). Where do the posterior tibial a and tibial n lie with respect to these tendons?

Mallet, malleable (literally, hammerable). The malleus is one of the auditory ossicles in the tympanic cavity. It is caused to vibrate by movements of the tympanic membrane. What muscle damps down movements of the malleus and what is its innervation?

Ma, mam, mam[m]a, mammary, mammal. All derive from this basic word-root and the child's first sounds "ma-ma"

The mammillary bodies are two small, breast-like protruberances on the brain base at the upper border of the pons. In front of them is the pituitary stalk or infundibulum

The mandible is the bone of the lower jaw – the jaw that is free to move during chewing! What muscles are required for this action?

A salivary gland situated below the mandible. What is its secretomotor innervation?

Anatomical Names	Latin/Greek or Other Origins	Examples
Manubrium	**Manubrium**: [L] *hilt, handle* **manus**: [L] *hand*	Manubrium
Masseter, masseteric	**Masein**: [Gr] *to chew*	Masseter m Masseteric n
Mastication	**Masticare**: [L] *to chew*	Muscles of mastication
Mastoid	**Mastos**: [Gr] *breast*; **eidos**: [Gr] *shape*	Mastoid process
Mater, matrix	**Mater**: [L] *mother* (hence *protector* connotation) **matrix**: [L] *a female kept for breeding* (hence, *womb*)	Dura, arachnoid, pia mater
Maxilla, maxillae, maxillary	**Maxilla**: [L] *jawbone, jaw*	Maxilla Maxillary n
Meatus, meatuses	**Meatus**: [L] *way, channel*	External auditory meatus Nasal meatuses
Medial, median	**Medius**: [L] *middle, in the middle, middling*	Medial

Notes, Links and Non-Anatomical Usages

Short for manubrium sterni (the handle of the sternum). From the **manus** root, we also derive manual, manicure, manifesto, manufacture (hand-made) and manuscript (handwritten)

The masseter is a muscle of mastication (chewing)

Supplies the muscle and arises from which division of cranial V?

Masticatory. Mastic is an aromatic gum from the mastic tree and is used to flavour chewing gum!

Mastitis, mastectomy, mastodon (breast-like tooth – an extinct elephant-like mammal with nipple-like tubercles on the crowns of its molar teeth). The mastoid process is part of the temporal bone and, after about the first postnatal year, is a breast-like projection lying posterior to the internal acoustic meatus. Why is it poorly developed in neonates?

Mater (Victorian term for mother; the meninges help to protect the brain and spinal cord), maternal, matron, matriculate, matrimony. A matrix may also be supportive or nutritive (like the womb), e.g. extracellular matrix

The upper jawbone. The bone is less dense than that of the mandible making it easier for dentists to anaesthetise the teeth and gums simultaneously. In the mandible, teeth and gums must be anaesthetised by separate injections

This nerve is a branch of cranial V. It is exclusively sensory. What is its area of supply?

The external auditory (or acoustic) meatus is the channel running from the auricle to the tympanic membrane. It slopes downwards anteromedially and contains ceruminous glands secreting wax. How might this wax (or other irritation of the mucosa) trigger a cough reflex?

The superior, middle and inferior meatuses are channels running anteroposteriorly between the nasal conchae and floor of the nasal cavity. What structures open into each meatus?

Medium, medieval, mediocre (literally, half way up the mountain, but meaning of moderate quality!). Medial structures lie closer to the body's midplane. A median structure lies in the midplane

Anatomical Names	Latin/Greek or Other Origins	Examples
		Median n
Mediastinum	**Mediastinum**: [L] *a menial (slave), his/her quarters*	Mediastinum
Medulla, medullary, medullaris	**Medulla**: [L] *marrow, pith, inmost part* (related to medius and can imply *'in the midline'*)	Adrenal medulla
		Medulla oblongata
		Conus medullaris
Meninx, meninges, meningeal	**Meninx**: [Gr] *membrane*	The meninges
		Middle meningeal a
Meniscus, menisci	**Meniskos**: [Gr] *crescent, little moon*	Medial and lateral meniscus
Mentum, mental	**Mentum**: [L] *chin*	Mental n
Mesentery, mesenteric	**Mesos**: [Gr] *middle;* **enteron**: [Gr] *gut, intestine*	The mesentery

Notes, Links and Non-Anatomical Usages

The median n of the upper limb arises from the lateral cord of the brachial plexus and runs roughly down the middle of the anterior compartments of the upper limb. It supplies muscles in the anterior compartments of the forearm and hand. What is its sensory innervation?

The mediastinum is a set of regions into which the thoracic cavity can be divided. What are the different regions called?

The adrenal medulla is the inmost part of the gland

The medulla oblongata is in the midline

The medullary cone. Here medullary refers to the spinal medulla (or spinal cord), the distal end of which is cone-shaped

The meninges are a set of three membranes surrounding the brain and spinal cord (dura, arachnoid and pia mater)

The largest of the meningeal aa. It enters the cranial cavity through which foramen?

The menisci (or semilunar cartilages!) of the knee joint are compressed fibrous discs attached to the superior surface of the tibia. The medial meniscus is C-shaped, larger and less mobile (why?). The lateral one is almost circular, smaller and more mobile.

Not to be confused with mental from **mens**: [L] *mind*. The mental foramen of the mandible transmits which nerve? Of what nerve is it a terminal branch?

The mesentery slings the abdominal foregut to the posterior abdominal wall and is found in the middle of the intestinal mass

97

Anatomical Names	Latin/Greek or Other Origins	Examples
		Superior mesenteric a
Mitral	**Mitra**: [L] *type of cap, turban*	Mitral (bicuspid) valve
Modiolus	**Modiolus**: [L] *nave or hub of a wheel*	Modiolus (of cochlea or angle of mouth)
Molar	**Molarius**: [L] *concerned with milling or grinding* **molaris**: [L] *related to a mill-stone*	Molar teeth
Mons	**Mons**: [L] *mountain*	Mons pubis
Mucosa(e), mucosal, mucous	**Mucosus**: [L] *slimy, mucous*	Mucosa
Muscle, muscular[is]	**Musculus**: [L] *little mouse, muscle*	Muscle
		Muscularis mucosae

N

Nares	**Naris**: [L] *nostril*	Nares
Nasion, naso-	**Nasus**: [L] *nose*	Nasion

Notes, Links and Non-Anatomical Usages

The arterial supply of the midgut

This left atrioventricular valve has two tapered cusps and resembles the mitre, headwear worn by bishops. The mitre has two tapering points at the front and back, separated by a cleft

The term is used to define either the central conical axis of the cochlea of the ear (modiolus cochleae) which contains the spiral ganglion or the fibromuscular mass at the angle of the mouth into which various muscles of facial expression (the modiolar mm) are inserted. In both cases, the meaning of a centre or hub is intended

Molar (a tooth for grinding food)

Monte Carlo, Montreal. In females, the mons pubis is not so much a mountain as a mound – the mons veneris (or Mound of Venus). This is a rounded eminence lying anterior to the pubic symphysis and formed by hairy skin overlying a mass of subcutaneous adipose tissue. It comprises the two labia majora separated by the pudendal cleft

Mucus, mucin, mucopolysaccharide. A mucosa is a mucous membrane lining several hollow and tubular organs such as the intestines, uterus and respiratory passages

Muscular. Muscle was named, by analogy, to the jerky, darting movements of a mouse

The smooth muscle layer associated with a mucosa (e.g. of the small intestine)

The nostrils are separated by the nasal septum and may be widened to control the intake of air

Nasal. The nasion lies at the bridge of the nose where the two nasal bones meet the frontal bone. It is one of multiple sites used for making cranial measurements

Anatomical Names	Latin/Greek or Other Origins	Examples
		Nasopharynx
Navicular	**Navicula**: [L] *little boat*	Navicular bone
Nephron, nephric	**Nephros**: [Gr] *kidney*	Nephron
		Mesonephric duct
Node, nodular	**Nodus**: [L] *knot, tie, swelling*	Lymph node
		Flocculonodular lobe
Nucha, nuchae, nuchal	**Nucha**: [ML] *medulla oblongata* and **nuka**: [Arab] *spinal medulla*	Ligamentum nuchae or nuchal lig
		Superior nuchal lines
Nucleus	**Nucleus**: [L] *kernel, nut, inside*	Nucleus pulposus

Notes, Links and Non-Anatomical Usages

The nasal part of the pharynx, posterior to the nasal cavities. Its roof and posterior wall lie inferior to the body of the sphenoid bone and basilar part of the occipital bone

Navy, naval, navigation. The disarticulated bone resembles a little boat

Nephric (meaning the same as renal), nephrology, nephritis, nephropathy, epinephrine (= adrenaline). The nephron is the main functional unit of the kidney, comprising the renal corpuscle and its associated uriniferous tubule

During embryological development three kidney systems develop sequentially, the middle of which is the mesonephros (middle kidney). The excretory tubules first appear in the mesonephros. At one end, they form the renal corpuscle and, at the other, they enter the collecting (mesonephric or Wolffian) duct

Nodule, nodose

Phylogenetically, an ancient part of the cerebellum. See Flocculus above

Nowadays, nucha refers to the back or nape of the neck. The nuchal lig is not a true ligament but a dense fibro-elastic band extending from the external occipital protuberance, via the tips of cervical spinous processes, to the spinous process of the 7th cervical vertebra. Superficially, this corresponds to a shallow midline groove (the nuchal groove) running down the back of the neck

These are bony ridges running laterally across the occipital bone from the external occipital protuberance. Together with the nuchal lig, they provide attachments for which important muscle?

Nuclear, nucleated, nucleolus, nucleic acid, nuclease, etc. The nucleus pulposus is the elastic, central part of an intervertebral disc. Inordinate pressure may cause this to be squeezed through the outer annulus fibrosus to produce a ruptured or herniated disc. This may impinge on the spinal cord causing pain, numbness or loss of muscle function

Anatomical Names	Latin/Greek or Other Origins	Examples
O		
Obturator	**Obturare**: [L] *to close, stop up, block off, obstruct*	Obturator foramen
Occiput, occipital	**Occiput**: [L] *back of head*	Occiput
		Occipital condyle
Ocular, oculi, oculo-	**Oculus**: [L] *eye*	Extraocular mm
		Orbicularis oculi m
		Oculomotor n (cranial III)
Odontoid, -odontal	**Odon**: [Gr] *tooth*; **eidos**: [Gr] *shape*	Odontoid process
		Periodontal lig
Olecranon	**Olene**: [Gr] *elbow*; **kranion**: [Gr] *head*	Olecranon
Olfaction, olfactory	**Olfactus**: [L] *odour, sense of smell*	Olfaction
		Olfactory n (cranial I)

Notes, Links and Non-Anatomical Usages

To obturate (to obstruct). The obturator foramen, in life, is partially closed by the obturator membrane through which passes the obturator n

The occipital bone articulates with the atlas and has some important openings: the foramen magnum and anterior condylar (or hypoglossal) canal

The occipital condyles articulate with superior facets on the atlas and these atlanto-occipital joints allow what movements?

Ocular (also used to describe the eyepiece of a microscope), oculist. The extraocular (outside the eye) mm are the extrinsic muscles which move the eye within the orbit. What are their names, actions and innervations?

The orbicularis oculi m (encircling muscle of the eye) is a muscle of facial expression allowing you to close the eyelids and 'screw up' your eyes. As you know, it is supplied by the facial (cranial VII) n

The oculomotor (eye-moving) m supplies which extraocular mm?

Odontology, orthodontist (makes sure you have 'straight teeth'). The odontoid process is shaped like a tooth and its alternative name (**dens**: [L] *tooth*) proves it!

The periodontal lig, or periodontal membrane, 'surrounds the tooth'. It is a dense fibrous connective tissue supporting the tooth in its socket and lying between the alveolar bone and the dental cement

The head of the elbow – actually the bony prominence at the proximal end of the ulna

Olfaction is the special sense of smell

Helps to connect the olfactory mucosa of the upper nasal cavity to the olfactory cortex ('smell brain')

Anatomical Names	Latin/Greek or Other Origins	Examples
Omentum, omental	**Omentum**: [L] *bowels, fat, membrane, caul*	Greater omentum
		Omental bursa
Omohyoid	**Omos**: [Gr] *humerus, shoulder* (see Hyoid above)	Omohyoid m
Ophthalmic	**Ophthalmos**: [Gr] *eye*	Ophthalmic a
Opponens	**Opponens**: [L] *setting against*	Opponens pollicis m
Optic	**Optikos**: [Gr] *related to sight*	Optic n
Oral, oris, oro-	**Oralis**: [L] *of the mouth*	Oral cavity
		Orbicularis oris m
		Oropharynx
Orbicularis	**Orbiculus**: [L] *a little ring*	Orbicularis oculi and oris mm
Ostium	**Ostium**: [L] *door, entrance, opening*	Ostium
Otic, oto-	**Otikos**: [Gr] *related to the ear*	Otic ganglion
		Otocyst

Notes, Links and Non-Anatomical Usages

The greater omentum is suspended from the greater curvature of the stomach and is supplied by which arteries?

This lesser peritoneal sac lies posterior to the lesser omentum and communicates with the greater sac via the epiploic foramen

The omohyoid m (nerve supply?) joins the shoulder (actually the scapular notch on the superior scapular border) to the hyoid bone. Like the digastricus m, it has two bellies. Recall that the acr[o]-omion (acromion) forms the tip of the shoulder

Ophthalmology. The ophthalmic a arises from the internal carotid a in the middle cranial fossa and enters the orbit via which opening?

Opponent, opposite, opposition. What is opposition of the thumb? The muscle attaches to the trapezium and 1st metacarpal. It is supplied by the median n. Are there other opponens muscles?

Optical, optician. The optic n is cranial n II

Oration (what comes out of the mouth!)

A muscle of facial expression which compresses the lips

The mouth part of the pharynx, as distinct from the nasal (nasopharynx) and laryngeal (laryngopharynx) parts

Orbit (a circuit, sphere of activity or eye socket), orbicular (shaped like a ring or flat disc). Orbicularis oculi m surrounds the eye and is a sphincter of the eyelids. Orbicularis oris m surrounds the mouth and acts as its sphincter. What are their innervations?

A term referring to the openings of various structures including the maxillary sinus, uterine tubes, uterus and cervix

Otorhinolaryngology (ear, nose and throat), parotid (near the ear), otitis media (inflammation of the middle ear)

The inner ear rudiment

Anatomical Names	Latin/Greek or Other Origins	Examples
P		
Palatine, palati[ni], palato-	**Palatum**: [L] *palate*	Palate
		Palatine nn
		Levator veli palati[ni] m
		Palatopharyngeus m
Palpate, palpation	**Palpare**: [L] *to touch gently*	Palpation
Palpebra, palpebrae, palpebral	**Palpebra**: [L] *eyelid*	Palpebral fissure
		Levator palpebrae superioris m
Pampiniform	**Pampinus**: [L] *tendril, vine-shoot*	Pampiniform venous plexus
Papilla, papillae, papillary	**Papilla**: [L] *nipple, teat, breast*	Greater duodenal papilla
		Papillary mm
Parietal	**Paries**: [L] *wall*	Parietal (bones, peritoneum, pleura, etc)

Notes, Links and Non-Anatomical Usages

The palate comes hard or soft, like a boiled egg

Greater and lesser palatine nn and nasopalatine nn supply the hard and soft palates. From which cranial openings do they emerge?

Raises the soft palate during swallowing

A vertical pharyngeal muscle which also assists swallowing. How?

Palpate (to inspect anatomically by gentle touch), palpable (able to be felt or touched)

The eyelids (palpebrae) are connected to the medial and lateral margins of the orbit by palpebral ligs. The palpebral fissure is the line made when the eyelids are closed

As the name suggests, this muscle raises the upper eyelid. What is histologically unusual about this muscle? How is this reflected in its innervation?

The pampiniform plexus associated with the testicular blood supply is a complex network of veins which are tributaries of the testicular v and closely associated with the testicular a (like tendrils around the stem of a vine). There is countercurrent heat exchange between artery and venous plexus which helps maintain a lower testicular temperature

Papilloma (a wart-like growth). However, the colloquial term pap (teat, nipple, breast) is ON in origin and is probably imitative of the lip-smacking sounds at suckling. The greater duodenal papilla is where the bile and main pancreatic ducts converge to enter the 2nd part of the duodenum on its posteromedial wall

Extensions of myocardium into the ventricular cavities which help to close the atrioventricular valves

Parietal literally means related to the walls and all anatomical structures with this name are exactly that

Anatomical Names	Latin/Greek or Other Origins	Examples
Parotid	**Para-**: [Gr] *near;* **otos**: [Gr] *ear*	Parotid gland
Patella, patellae, patellar	**Patella**: [L] *small dish, plate, pan*	Patella Infrapatellar bursa
Pecten, pectineal, pectinati	**Pecten**: [L] *comb, rake*	Pecten pubis, pecten ani Pectineus m Musculi pectinati
Pectoral[is], pectoris	**Pectus**: [L] *breast (= chest)*	Pectoralis major m Clavipectoral fascia
Peduncle, pedunculi, peduncular	**Pedunculus**: [L] *small foot, stalk*	Cerebral peduncles Interpeduncular fossa

Notes, Links and Non-Anatomical Usages

Parotid (situated near the external ear). The gland lies anterior to the mastoid process, external ear and sternocleidomastoid m and passes superficial and deep to the ramus of the mandible. These, and other, salivary glands swell in mumps patients. Swelling of the parotid gland is particularly painful because its fibrous capsule is especially strong superficially

Hence, Spanish 'paella' which is cooked in a large shallow pan. The patella (bone of the knee-cap) resembles an inverted dish or small bowl

Subcutaneous and deep infrapatellar bursae are associated with the knee joint (patella and patellar lig). They may become enlarged after prolonged kneeling. Do they communicate with the knee joint cavity?

The symbol for the Shell Oil Company is a pectinate mollusc! Areas with this name usually have some comb-like appearance or relationship

A muscle of the thigh helping to form the floor of the femoral triangle. It arises from the pecten pubis of the superior pubic ramus and inserts just below the lesser trochanter of the femur. It is usually supplied by the femoral n but may also be supplied by which other nerve?

These muscles are found in the wall of the right atrium of the heart. The back of the comb is the crista terminalis and the teeth of the comb are the musculi pectinati

Angina pectoris (chest pain). The pectoralis major m has two heads: clavicular and sternocostal. These fibres collectively insert where?

Deep fascia investing the pectoral minor and subclavius mm and attaching to the axillary fascia and clavicle. The part between the upper border of pectoralis minor m and the clavicle is sometimes called the costocoracoid membrane and is pierced by the cephalic v, thoracoacromial a and lateral pectoral n

The pedunculate oak is the common or English oak. It produces stalked acorns. The cerebral peduncles join the cerebrum to the midbrain. Each has a ventral crus cerebri and dorsal tegmentum separated by substantia nigra

An area associated with the optic chiasma, tuber cinereum and pituitary stalk, mammillary bodies and posterior perforated substance. The crura cerebri of the cerebral peduncles form the caudolateral boundaries of this fossa

Anatomical Names	Latin/Greek or Other Origins	Examples
Pellucidum	**Pellucidus, perlucidus**: [L] *bright, transparent*	Septum pellucidum
Pelvis, pelves, pelvic	**Pelvis**: [L] *basin*	Pelvis
		Pelvic diaphragm
Penis, penile	**Penis**: [L] *tail*	Penis
		Penile urethra
Perineum, perineal	**Peri**-: [Gr] *around*; **naien**: [Gr] *to dwell*	Perineum
		Perineal membrane
Peritoneum, peritoneal	**Peri**-: [Gr] *around*; **tonos**: [Gr] *stretched*	Peritoneum
Peroneus, peroneal	**Perone**: [Gr] *fibula, pin*	Peroneus longus m
		Common peroneal n

Notes, Links and Non-Anatomical Usages

Pellucid (transparently clear). The septum pellucidum is a thin partition comprising two laminae each of which forms part of the wall of a lateral ventricle. The laminae are covered by ependyma laterally and by pia mater medially. The septum harbours septal nuclei and fibre connections

Pelvimetry (obstetrical measurement of pelvic dimensions). The pelvic region is made of the pelvic and sacral bones and is divided into an upper or false pelvis (false because this basin has no anterior wall) and a lower or true pelvis

The muscular sling for pelvic viscera comprising the levator ani and coccygeus mm

May derive from **pendere**: [L] *to hang* (as in depend). It can also mean 'to be flabby' and, certainly, it is flaccid when it is hanging down! However, remember that its anatomical position is erect and, in this state, it is turgid and its dorsal surface faces the anterior abdominal wall

The male urethra is divisible into different regions (what are they called?), the last of which is the penile or spongiose region which passes within the corpus spongiosum and extends from the perineal membrane to the external urethral orifice

Possibly from the same root as **perina**: [Gr] *scrotum*. Indeed, the term perineum originally covered the area between the anus and the scrotum or vulva. Anatomically, it defines the space between skeletal boundaries which are anterior (pubic arch and arcuate pubic ligs), posterior (tip of coccyx), lateral (inferior ischiopubic rami, ischial tuberosities and sacrotuberous ligs). A line running between the tuberosities divides the space into two triangles. What are their names?

A fibrous membrane forming the superior border of the superficial perineal pouch and the inferior border of the deep perineal pouch. It used to be referred to as the inferior fascia of the urogenital diaphragm

Peritoneum stretches around the abdominal cavity and viscera. Peritonitis (inflammation of peritoneum)

Peroneal means the same as fibular. This muscle is an evertor of the foot supplied by the superficial branch of the common peroneal n

The nerve passes into, and divides, on the lateral (fibular) side of the leg, into deep and superficial branches

Anatomical Names	Latin/Greek or Other Origins	Examples
Pes, pedis	**Pes**: [L] *foot*	Pes planus
		Dorsalis pedis a
Petrous, petrosal	**Petra**: [L] *rock, stone*	Petrous part of temporal bone
		Petrosal sinuses
Phalanx, phalanges, phalangeal	**Phalanx**: [Gr] *bone of finger or toe*	Proximal phalanx The phalanges
		Metacarpophalan-geal joint
Philtrum	**Philein**: [Gr] *to love*	Philtrum
Phrenic	**Phrenikos**: [Gr] *related to the diaphragm*	Phrenic n
		Musculophrenic a
Pia, pial	**Pia**: [L] *tender, delicate*	Pia mater

Notes, Links and Non-Anatomical Usages

Pedicle (little foot), pedal, pedestrian, pedicure, pedicel. Pes planus is flat foot

A lower limb pulse site on the dorsum of the foot. Situated where precisely?

Petrify (turn to stone), Peter the Rock (the most steadfast of the disciples of Jesus. Actually, the name Peter means 'the rock' anyway!). The petrous part of the temporal bone is its densest and helps to protect delicate internal structures like the inner ear

Dural venous sinuses which help drain the cavernous sinus

Phalangeal. In ancient Macedonia, Alexander the Great inherited from his father, Philip, a formidable fighting unit, a line of infantry in close ranks, called a phalanx. The thumb and big toe have two phalanges (proximal, distal) but all other fingers and toes have an additional intermediate or middle phalanx

These joints are articulations between metacarpal and proximal phalangeal bones

A love-philtre or love-potion is a drink exciting sexual passion in the drinker. The philtrum is the shallow vertical groove in the middle of the upper lip. What is the link to sexual passion? Could it be kissing? Or is it that the lateral columns of the philtrum cause the upper margin of the upper lip to resemble Cupid's bow?

Phrenic (in Greek, also relates to the mind, hence phrenology. This is because they thought that the mind resided in the diaphragm. No, I can't see the logic either!)

The internal thoracic a in the 6th intercostal space divides into two branches. The musculophrenic a is one and it supplies the periphery of the diaphragm and the adjacent thoracic wall. What is the other branch and what does it supply?

The innermost and most delicate of the meninges. It is the layer which accompanies small blood vessels as they enter the substance of the brain

Anatomical Names	Latin/Greek or Other Origins	Examples
Pinna, pinnae, pennate	**Penna, pinna**: [L] *feather, wing, fin*	Uni-, bi-, multi-pennate
		Pinna
Piriform	**Pirus**: [L] *pear*, **forma**: [L] *shape*	Piriformis m
Pisiform	**Pisum**: [L] *pea*; **forma**: [L] *shape*	Pisiform bone
Placenta	**Plakoeis**: [Gr] *flat cake;* **placenta**: [L] *pancake*	Placenta
Plantar, plantaris	**Planta**: [L] *sole of the foot*	Long plantar lig
		Plantaris m
Platysma	**Platys**: [Gr] *flat, wide, broad, ample*	Platysma m
Pleura, pleurae, pleural	**Pleura**: [Gr] *side or rib*	Parietal pleura
		Pleural cavity

Notes, Links and Non-Anatomical Usages

Pen (because pens were originally made from quills), penne (a type of pasta in the form of short wide tubes resembling the quill or hollow shaft of a feather). A bipennate muscle is like a feather: its tendon is the shaft and its fibres run into it like the vanes or barbs of the feather (e.g. dorsal interossei mm). A unipennate muscle has fibres running into the tendon from one side only (e.g. flexor pollicis longus m). A multipennate muscle may have multiple bipennate units running into each other (e.g. deltoid m)

Pinnate, pinnacle, pinniped (literally 'fin-foot' and applied to flippered mammals like seals, sea-lions, etc). The pinna or auricle is the external part of the ear

Piriform or pyriform (pear-shaped). Several anatomical or microscopic structures are described as piriform but the least convincing is the piriformis m! It laterally rotates the femur at the hip joint and can act as an abductor of the flexed femur. What are its attachments?

A pea-shaped carpal bone. The proper name for the garden pea is *Pisum sativum* (satisfying pea). The old word for pea, pease (as in the nursery rhyme "Pease pudding hot, pease pudding cold...etc"), is from the same word-root

Aptly named if you have seen one just delivered!

Plantigrade (walking on the sole of the foot, like humans). The long plantar lig strengthens the calcaneocuboid joint inferiorly

A vestigial muscle not always present. It is associated with the lateral head of gastrocnemius m and inserts on the medial side of the posterior part of calcaneus. What do you think would be its motor nerve supply?

Plato (the Greek philosopher was broad-shouldered), platyhelminths (flatworms), platypus (broad of foot). The platysma m is subcutaneous and a muscle of facial expression used when grimacing. Nerve supply?

Parietal pleura lines the ribcage, pleurisy (inflammation of pleura)

The potential space between parietal and visceral pleurae

Anatomical Names	Latin/Greek or Other Origins	Examples
Plexus	**Plexus**: [L] *plaited, interwoven, braided*	Sacral plexus
Plica, plicae	**Plicare**: [L] *to fold*	Plicae circulares
Pollex, pollicis	**Pollex**: [L] *thumb*	Pollex
		Flexor pollicis longus m
Pons, pontine	**Pons**: [L] *bridge, drawbridge*	Pons
		Pontine cistern
Popliteus, popliteal	**Poples**: [L] *knee*	Popliteal fossa
		Popliteus m
Porta, portal, porto-	**Porta**: [L] *door, entrance, outlet*	Porta hepatis
		Portosystemic anastomosis
Profunda, profundus	**Profundus**: [L] *deep, profound*	Profunda femoris a

Notes, Links and Non-Anatomical Usages

Complex, perplex. Plexuses are formed from nerves or veins. The sacral plexus of nerves is formed by the lumbosacral trunk (L4, L5) and ventral rami of S1-S4. What are its branches? A sacral venous plexus joins lateral sacral vv (tributaries of the internal iliac vv) which accompany the lateral sacral aa

Plicate (folded), pliers (for bending wire), plywood (made up of layers). Plicae circulares are folds of mucosa running round the circumference of the small intestine. Some, but not all, are complete circles

Pollex, in Latin, also meant big toe which is now called the hallux

The long flexor of the thumb is supplied by which nerve?

The pons bridges the brain hemispheres. Pontoon bridge! Pontefract (broken bridge) in Yorkshire is the traditional site of UK manufacture of liquorice sweets. Pont l'Evêque (bishop's bridge) in Normandy is the site of manufacture of a favourite cheese of mine

Contains cerebrospinal fluid and lies anterior to the pons. The basilar a runs through the cistern in the midline

The popliteal fossa lies behind the knee joint. What are its boundaries?

The muscle forms part of the floor of the popliteal fossa. It arises from the lateral femoral condyle and the arcuate popliteal lig and inserts on the tibia above the soleal line. It unlocks the knee prior to flexion. What movement does unlocking involve?

Airport, portal (door in poetry), portcullis (sliding door), portico. The porta hepatis (portal area of the liver) is the hilum where vessels enter (hepatic a, portal v) and leave (bile duct)

Occurs between portal and systemic venous circulations. In portal hypertension, where are the commonest sites of venous distension for diagnostic purposes?

Profound (spatially or intellectually deep). Remember that deep in Anatomy always means further away from the skin surface. The profunda femoris a is a deep muscular branch of the femoral a. Its own branches contribute to which important anastomoses?

Anatomical Names	Latin/Greek or Other Origins	Examples
		Flexor digitorum profundus m
Pronation, pronator	**Pronus**: [L] *bent, inclined towards, bowing*	Pronation
		Pronator teres m
Protraction, protractor	**Pro-**: [L] *forwards*; **trahere**: [L] *to draw, pull*	Protraction
Protrusion, protrusor	**Pro-**: [L] *forwards*; **trudere**: [L] *to thrust*	Protrusion
Proximal	**Proxime**: [L] *nearest, next, very close to*	Proximal
Psoas	**Psoa[i]**: [Gr] *loin muscle*	Psoas major and minor mm
Pterion	**Pteron**: [Gr] *wing, anything wing-like*	Pterion
Pterygoid, pterygo-	**Pteron**: [Gr] *wing, anything wing-like*	Lateral pterygoid plate

Notes, Links and Non-Anatomical Usages

The deep flexor of the fingers. At what joints does this muscle produce flexion?

Prone (meaning lying face down or inclined to), pronate. Pronation in the forearm involves rotation of the radius and its movement relative to the ulna so that the palm of the hand faces posteriorly. The radius turns anteromedially to lie obliquely across the ulna. Which movement is stronger, pronation or supination?

One of the two main pronators of the forearm, the other being pronator quadratus m. Both are supplied by the median n. What other forearm muscle may assist in pronation/supination?

Protracted, extract, tractor. In protraction of the upper limb (as when extending the reach), the scapula is pulled anteriorly. By what muscles?

Protrude. Protrusion of the mandible is brought about by the pterygoid mm, particularly the lateral pterygoid mm. These are supplied by the mandibular division of cranial n V

Approximate (close to), proximity. In a limb, proximal structures are closer to the root of the limb. Proxima Centauri is the closest star to our Solar system

The psoas major m attaches to lumbar vertebrae and the lesser trochanter of the femur. It is innervated by ventral rami of L1-L3 and flexes the thigh or trunk at the hip joint. Less than 50% of people have a psoas minor m. What are its attachments, innervations and action? Psoas major is one of the muscles in a pork, lamb or beef chop but psoas mm are the only muscles in the cut known as tenderloin

Pterodactyl (a reptile with wings supported by an elongated digit), pterygotes (the winged insects). The pterion is on the lateral aspect of the skull and represents the smallest circle which contains the suture at which the frontal, parietal, sphenoid and squamous temporal bones meet. The name may derive from the fact that the sphenoid component is from the greater wing of the sphenoid. The bone at this site is liable to fracture because it is naturally thin and, in addition, eroded internally by which artery?

The pterygoid plates are wing-like postero-inferior extensions of the sphenoid bone which, itself, has lesser and greater wings! The lateral plate provides origins for both the lateral and medial pterygoid mm

Anatomical Names	Latin/Greek or Other Origins	Examples
		Pterygomaxillary fissure
Ptosis	**Ptosis**: [Gr] *drooping, falling*	Ptosis
Pudenda, pudendal	**Pudendus**: [L] *shameful, disgraceful*	Pudenda
		Pudendal n
Pulmonary	**Pulmo**: [L] *lung*	Pulmonary a
Pulvinar	**Pulvinar**: [L] *sofa, bed, couch (all cushioned)*	Pulvinar
Putamen	**Putamen**: [L] *shell (of fruit), husk, peelings, clippings, waste*	Putamen
Pylorus, pyloric	**Pyloros**; [Gr] *gatekeeper*	Pylorus Pyloric sphincter
Q		
Quadrate, quadratus	**Quadratus**: [L] *square, quadrangular*	Quadrate tubercle Quadratus femoris m

Notes, Links and Non-Anatomical Usages

This is the entrance to the pterygopalatine fossa. It lies between infratemporal surface of the maxilla and the lateral pterygoid plate. What are the boundaries and contents of the fossa?

Ptosis of the upper eyelid is a diagnostic sign of what? (Name the muscles and nerves). Apoptosis (falling away) is loss of cells by programmed cell death

Pudenda (the external genitalia - short for pudenda membra meaning, literally, the parts to be ashamed of! We still speak of them as the "naughty bits"). An impudent person originally was one who had no shame or displayed his/her genitals in public – a streaker?

One of the main terminal branches of the sacral plexus (S2-S4). It runs between the piriformis and coccygeus mm to enter the greater sciatic foramen. It then passes behind the sacrospinous lig and into the lesser sciatic foramen, thereby bypassing the pelvic diaphragm. It enters the pudendal canal, a fascial sheath on the lateral wall of the ischiorectal fossa. What does the nerve supply?

Pulmonates (molluscs, including slugs and snails, in which the mantle cavity is modified to act as a lung). Remember that not all arteries convey oxygenated blood. The pulmonary aa convey deoxygenated blood from the right ventricle to the lungs

The pulvinar is so-called because this part of the thalamus projects like a cushion or knob

Part of the lentiform nucleus. Maybe named because, in a frontal slice, the whole lentiform nucleus looks like odd scraps?

The pyloric sphincter guards the pylorus (gate between stomach and duodenum)

Quadrat. The quadrate tubercle of the intertrochanteric crest of the femur provides attachment for the quadratus femoris m. What are the actions of this muscle? Also, the liver has a quadrate lobe

Anatomical Names	Latin/Greek or Other Origins	Examples
		Quadrangular space
Quadriceps	**Quadriceps**: [L] *four-headed*	Quadriceps femoris m
R		
Radius, radii, radial[is]	**Radius**: [L] *stick, rod, poke, weaver's shuttle*	Radius
		Radial a
		Flexor carpi radialis m
Ramus, rami	**Ramus**: [L] *branch, twig*	Ramus of the mandible
		Rami communicantes
Raphe	**Raphe**: [Gr] *stitching, seam*	Raphe of the pharynx
Rectum, rectus, recto-	**Rectus**: [L] *straight, regular*	Rectum

Notes, Links and Non-Anatomical Usages

The quandrangular (four angles) or quadrilateral (four sides) space forms part of the course of the axillary n. The space is bounded above by the teres minor and subscapularis mm, below by teres major m, laterally by the surgical neck of the humerus and medially by what? Into which branches does the axillary n divide and what do the branches supply?

Of course, you know that the four heads are rectus femoris, vastus lateralis, medius and intermedius mm

Radius (rod or stick is not a bad description of the bone), radial. The bone is on the lateral side of the forearm and moves with respect to the ulna during pronation and supination

Arises from the brachial a in the cubital fossa. Its pulse may be palpated where?

The flexor of the wrist on the radial (= lateral) side. It arises from the common flexor origin on the medial epicondyle of the humerus and inserts on the palmar aspect of metacarpal bases 2 and 3. It is supplied by the median n. What are its antagonists and synergists in movements at the wrist?

Ramify (to form branches), ramification. The ramus of the mandible extends from the angle and runs upwards to branch into the condylar and coronoid processes

Ventral rami of spinal nn T12-L2 have branches called white rami communicantes which pass to ganglia on the sympathetic trunk. Axons from ganglion cells pass from the trunk and rejoin the spinal nn as grey rami communicantes. Fibres in white rami are myelinated and preganglionic. Those in grey rami are non-myelinated and postganglionic. To what structures are the postganglionic sympathetic fibres in the spinal nn distributed?

A raphe tends to be a midline seam where sheets of tissue (usually muscle) meet. That of the pharynx provides attachment for the constrictor mm

Rectum (hopefully, you stay regular!), rectal, rectangular (right-angled), rectify (to put straight), rectilinear (in a straight line), rectitude

Anatomical Names	Latin/Greek or Other Origins	Examples
		Rectus abdominis m
		Rectouterine pouch
Recurrent	**Recurrere**: [L] *to return, run back*	Recurrent laryngeal n
Renal	**Renes**: [L] *kidneys*	Renal vv
		Adrenal (suprarenal) gland
Rete	**Rete**: [L] *net, fishing-net*	Rete testis
Reticulum, reticular	**Reticulum**: [L] *little net*	Endoplasmic reticulum
		Reticular formation

Notes, Links and Non-Anatomical Usages

The 'straight muscle of the abdomen'. The strap-like muscles on either side run down the anterior abdominal wall from lower thoracic cage to pubis and are separated by the linea alba. They are supplied by ventral rami of T7-T12 spinal nn and enclosed in the rectus sheath. What are the actions of the muscle and the contributors to the anterior and posterior parts of the rectus sheath?

The rectouterine pouch (of Douglas) is a peritoneal reflection from the posterior of the uterus on to the rectum

The vagus n descends from the brain into the thorax where it gives rise to a recurrent laryngeal branch which turns back and ascends to the larynx. Imagine how long this pathway is in a giraffe! What thoracic structures do the left and right recurrent laryngeal nn curve round before they ascend?

Adrenal, adrenaline, suprarenal, renin. The renal vv drain into the IVC but are of unequal length. On which side is the renal v shorter?

Adrenal means 'near the kidney' and, interestingly, American English borrowed the Greek for this (epinephron) in order to derive epinephrine which is the American equivalent of adrenaline

Rete (used to describe complexes of small vessels), retiarius (a type of Roman gladiator who fought with a net and trident), retina. The rete testis is a network of anastomosing tubes lying between the seminiferous tubules of the testis and the efferent tubules of the caput epididymis

Reticle, reticule (once a woman's bag or purse), reticulocyte (immature red blood cell with vestiges of protein synthetic machinery stainable with basic dyes and appearing as a reticulum). Endoplasmic reticulum is a subcellular organelle of interconnected tubules and cisternae involved in protein synthesis and export

A complex network of nerve fibres and grey matter scattered throughout the medulla oblongata, pons and midbrain. Sometimes called the reticular activating system, it is concerned with stimulating the cerebral cortex into a state of arousal or wakefulness. Decreased activity results in sleep. I hope that your reticular formation is currently active!

Anatomical Names	Latin/Greek or Other Origins	Examples
Retinaculum	**Retinaculum**: [L] *stay, tie*	Flexor retinaculum
Retraction, retractor	**Retro-**: [L] *backwards*; **trahere**: [L] *to draw, pull*	Retraction
Rhin-, Rhino-	**Rhis**: [Gr] *nose*	Rhinencephalon
Rhomboid	**Rhomboid**: [Gr] *rhombus-like in shape*	Rhomboid major m
Rima	**Rima**: [L] *crack, chink, fissure*	Rima glottidis
Risorius	**Risor**: [L] *one who laughs, scoffs or derides*	Risorius muscle
Rostrum, rostral	**Rostrum**: [L] *bill, beak, snout, anything projecting*	Rostrum of the corpus callosum
Rotation, rotator	**Rotare**: [L] *to turn (in a circle)*	Rotation (medial, lateral)

Notes, Links and Non-Anatomical Usages

A retinaculum retains tendons, or other structures, in place. The flexor retinaculum of the wrist constrains flexor tendons and stops them bow-stringing during wrist flexion. What other structures pass superior and deep to this retinaculum? This retinaculum is attached medially to the hook of the hamate (remember HH) and laterally to the tubercle of the trapezium (remember TT)

Retract (to withdraw). Latissimus dorsi m is a powerful retractor of the humerus. What muscles retract the scapula?

Rhinoceros (horny nosed), rhinitis (inflammation of the nasal mucosa), otorhinolaryngologist (ear, nose and throat specialist). The rhinencephalon comprises those regions of the cerebrum concerned with olfaction

A rhombus is a four-sided, oblique-angled parallelogram with equal sides. In a rhomboid, adjacent sides are of unequal length . The rhomboid major mm are rhomboidal muscles which help brace the shoulders by retracting the scapulae. They are supplied by the dorsal scapular nn but what are their attachments?

The rima glottidis is the gap between the vocal cords. The glottidis element refers to the tongue but, anatomically, strictly to the area of the vocal cords. Rima [L] is not to be confused with **rima**: [OE] *rim, border, edge*

Derisory, derision, ridicule. The risorius is a muscle of facial expression so what is its nerve supply? It is highly variable in form but runs between the parotid fascia to the skin at the modiolus (which, here, refers to the angle of the mouth). It derives its name from the fact that one of its actions is to pull the mouth into a 'false' or sardonic smile rather than a 'genuine' smile (which involves also zygomaticus and orbicularis oculi mm)

Rostrum, rostra (a platform for a speaker, conductor – in Roman times, the public platform was decorated with the ramming beaks from defeated enemy ships!). The rostrum of the corpus callosum projects anteriorly to the genu and downwards to the lamina terminalis (anterior wall of 3rd ventricle). Rostral (like cephalic) is also used to indicate nearer to the head, as opposed to the tail (caudal), end of the body

Rotate, rotary, rotor, rotifer (a multicellular invertebrate with a ciliated wheel-like organ used for feeding and locomotion). Recall that the axis of rotation of a long bone does not always correspond to the long axis of the bone. For instance, think of the femur

Anatomical Names	Latin/Greek or Other Origins	Examples
		Rotator cuff mm
Ruga, rugae	**Ruga**: [L] *wrinkle, crease*	Rugae (of bladder, scrotum, stomach, vagina)

S

Anatomical Names	Latin/Greek or Other Origins	Examples
Saccule, sacculus, sacculi, sacculation	**Sacculus**: [L] *little sac*	Saccule or sacculus
		Sacculation
Sacrum, sacra, sacral, sacro-	**Sacrum**: [L] *sacred thing, solemn site, sacrifice*	Sacrum
		Sacral promontory
		Sacroiliac joint
Sagittal	**Sagitta**: [L] *arrow*	Sagittal suture
Salpinx, salpingo-	**Salpinx**: [Gr] *tube, trumpet*	Salpinx
		Mesosalpinx
		Salpingopharyn-geus m

Notes, Links and Non-Anatomical Usages

Muscles (subscapularis, supraspinatus, infraspinatus, teres major) whose tendons help stabilise the shoulder joint

Rugose (wrinkled), corrugated. The mucosa of certain organs is thrown into ridges which create a wrinkled or corrugated appearance. The skin of the scrotum is also rugose. That explains why the testes are sometimes referred to as prunes!

The term sacculus is used to describe various anatomical sac-like spaces in the internal ear, larynx and elsewhere

In the large intestine, the term sacculation is sometimes used instead of haustration

Short for os sacrum (holy bone, because the Greeks and Romans thought that it housed the soul!), sacred, sacrifice

The sacral promontory is the anterior projecting rim of the upper surface of the first sacral vertebra. In the anatomical position, the centre of gravity of the body lies just below this promontory

A synovial joint between the sacral and iliac articular surfaces. In infants, the joint is plane but, in adults, movement is restricted by interlocking ridges and valleys. The range of movement increases in pregnant females

Sagittarius (the Archer constellation). A sagittal plane runs vertically through the body in the midline and in an anteroposterior direction. Hence the name sagittal suture

Salpiglossis (a plant with trumpet-like flowers), The term salpinx may refer to the uterine tube which, when blocked, requires surgical intervention, salpingostomy

The mesosalpinx is the mesentery of the uterine tube

Here, salpinx refers to the auditory tube. The muscle is supplied by the pharyngeal plexus but what are its attachments and actions?

Anatomical Names	Latin/Greek or Other Origins	Examples
Saphenous	Origin obscure: probably **saphena**:[ML] *vein*, or **saphene**: [Gr] *clear, obvious, plain*, but possibly **safin**: [Arab] *hidden, silent, secret!*	Great saphenous v
Sartorius	**Sartor**: [L] *tailor*	Sartorius m
Scala	**Scala**: [L] *ladder, staircase, spiral*	Scala tympani, media, vestibuli
Scalene, scalenus	**Skalenos**: [Gr] *unequal*	Scalene tubercle
		Scalenus mm
Scaphoid	**Scaphoid**: [Gr] *keel-shaped*	Scaphoid fossa, bone
Sciatic, ischium, ischial, ischio-	**Ischion**: [Gr] *hip*	Sciatic n
		Greater sciatic foramen
		Ischial tuberosity
		Ischiocavernosus m

Notes, Links and Non-Anatomical Usages

So, possibly, the great venous vein or obvious vein (when it is varicose?) or hidden vein (because it lies medially and, normally, is not varicose?)

Sartorial (related to a tailor or to tailoring). Traditionally, a tailor sat on a floor or bench with legs akimbo, i.e. with lower limbs flexed at hip and knee, and thigh abducted and laterally rotated. All actions of sartorius m!

Scalable (able to be climbed, like a ladder or staircase), scalariform cells (water-conveying plant cells with thickened, ladder-like walls), scale (a measuring scale or ruler is analogous to a ladder). The scala tympani (lower), media (middle) and vestibuli (upper) are spaces within the spiral cochlea of the inner ear

A scalene triangle has sides of unequal length. The scalene tubercle is a small projection on the superior surface of the 1st rib providing attachment for the scalenus anterior m. What lies in the groove anterior to the muscle? What lies in the posterior groove?

Scalene mm arise from cervical transverse processes and descend to attach to one or both of the first two ribs

At the base of the medial pterygoid plate is a hollow just like the keel of a boat! The carpal bone also has boat-like features

Sciatica (pain associated with the nerve). The sciatic n is a major terminal branch of the sacral plexus. What is the link between this nerve and foot-drop?

Lies above the level of the ischial spine and sacrospinous lig. What structures pass through it?

The ischial tuberosity is a roughened area of bone on the ischium which provides attachment for several muscles. Which ones?

This is one of them, a perineal muscle of the superficial perineal pouch which runs from the ischial tuberosity to the crus penis (male) or crus clitoridis (female). It helps compress the crus and so maintain erection of the penis or clitoris. The muscle is supplied by which branch of the pudendal n?

Anatomical Names	Latin/Greek or Other Origins	Examples
Sclera, sclerae, scleral, sclerous	**Skleros**: [Gr] *hard, tough*	Sclera
Sella	**Sella**: [L] *saddle*	Sella turcica
Septum, septa, septal, septo-	**Septum**: [L] *fence, enclosure*	Nasal septum Septal cartilages Septomarginal trabecula
Serosa, serous	**Serum**: [L] *whey, serum*	Serous membrane
Serratus	**Serratus**: [L] *like the teeth of a saw*	Serratus anterior m
Sesamoid	**Sesamon**: [Gr] *sesame, sesame seed*	Sesamoid bones
Sigmoid, sigmoidal	**Sigmoid**: [Gr] *S-shaped* (the equivalent letter S is called sigma in Greek)	Sigmoid sinus, colon
Sinus, sinu-	**Sinus**: [L] *hollow, cavity*	Oblique sinus Paranasal air sinuses Sinuatrial node

Notes, Links and Non-Anatomical Usages

The dense fibrous coat of the eyeball continuous anteriorly with the transparent cornea. Sclerosis, sclerite (a plate-like unit of the exoskeleton of arthropods)

The 'turkish saddle' (on account of its shape) or hypophyseal fossa (because it contains the hypophysis or pituitary gland) is a depression in the sphenoid bone lying above the sphenoidal air sinus

Septal cartilages and bone help to make the nasal septum which separates the 2 nasal cavities

A band of muscular and conducting tissue running across the right ventricle from the interventricular septum to the base of the anterior papillary m

A serosa (serous membrane) is a thin, moist membrane lining a cavity and partly or completely covering its viscera. Examples are the pleural and peritoneal membranes

Serratus anterior inserts on ribs 2-9 and interdigitates with attachments of external oblique m. The interdigitations look serrated, just like the teeth of a saw!

Sesamoid bones are 'seeded' and grow within tendons (e.g. the pisiform bone in flexor carpi ulnaris tendon and the patella in quadriceps femoris tendon)

Sigma, sigmoidal. All sigmoidal structures display an S-shape or a sinuous course. A sigmoidoscope is used to view internally the sigmoid colon

Sinusitis, sinusoid (sinus-like). The oblique sinus of the heart is a J-shaped space behind the left atrium created by pericardial reflections. The limb of the J contains the IVC and right pulmonary vv whilst the foot contains the left pulmonary vv

A set of cavities within skull bones lined by respiratory mucosa and communicating with the nasal cavities. What are their names and sites of drainage into the nasal cavity?

The pacemaker of the heart initiates the cardiac cycle and sets the basic beat frequency. Its cells are situated in the wall of the right atrium near the SVC opening and adjacent crista terminalis

Anatomical Names	Latin/Greek or Other Origins	Examples
Skull	**Skalle**: [ON] *skull*; **skål**: [ON] *drinking bowl*	Skull
Soleus, soleal	**Solea**: [L] *sole of the foot*	Soleus m
		Soleal line
Somatic	**Soma**: [Gr] *the body*	Somatic
Sphenoid, sphenoidal, spheno-	**Sphen**: [Gr] *wedge*; **eidos**: [Gr] *shape, form*	Sphenoid bone
		Sphenoidal air sinus
		Sphenomandibu-lar lig
Spinal	**Spina**: [L] *thorn, spine, prickle, backbone*	Spinal cord (medulla)
Spinosum	**Spinosus**: [L] *thorny, prickly, spiny*	Foramen spinosum
Splanchnic	**Splanchna**: [Gr] *the entrails, innards*	Splanchnic nn
Splenius, splenium	**Splenion**: [Gr] *bandage*	Splenius capitis and cervicis mm
		Splenium

Notes, Links and Non-Anatomical Usages

When Norsemen drink, they say "skål" which means "cheers". It may be that the drinking bowl was made originally from a skullcap of some animal. Is it just chance that people who drink to excess are described as being 'out of their skulls'?

This calf muscle which runs down towards the sole of the foot (actually inserting on the calcaneus) produces plantar flexion at the ankle

A line of attachment of the soleus m to the superior part of the posterior aspect of the tibia

Somatotype (classification of body type by physique or build). In Anatomy, somatic is used in two main senses: to contrast with visceral (related to organs) or with germinal (related to germ cells). For example, cranial nn contain four main types of nerve fibre: somatic afferent, visceral afferent, somatic efferent and visceral efferent

The bone is a complicated structure resembling two pairs of wings. It is wedged into the skull by surrounding bones. Which ones?

Opens and drains where?

A rather vestigial structure extending between which particular bits of the sphenoid bone and mandible?

Spine. The spinal cord does not extend the whole length of the vertebral column. In adults, it terminates at what level?

Spinose (= spiny). The foramen spinosum is just anteromedial to the spine of the sphenoid bone. What artery does it transmit?

Splanchnic sometimes means the same as visceral but refers to the organs in the abdomen rather than in general. There are sympathetic splanchnic nn (associated with thoracic spinal levels T5-T12) and parasympathetic splanchnic nn (sacral levels S2-S4). What viscera do they innervate?

These muscles help to extend the head (acting together) and rotate it (acting separately). In rotation, the splenii mm on one side act synergistically with the contralateral sternocleidomastoid m

What is the splenium?

Anatomical Names	Latin/Greek or Other Origins	Examples
Spongiosum	**Spongia**: [L] *sponge*	Corpus spongiosum
Stapes, stapedius	**Stapes**: [L] *stirrup*	Stapes
		Stapedius m
Stellate	**Stella**: [L] *star*	Stellate ganglion
Stratum, strata	**Stratum**: [L] *coverlet, horse blanket*	Stratum germinativum
Stria, striae, striate[d], striatum	**Stria**: [L] *groove, stripe*	Olfactory stria
		Striate cortex
		Striated muscle
		Corpus striatum
Stroma, stromal	**Stroma**: [Gr] *couch, bed, mattress, layer*	Stroma
Styloid, stylo-	**Stylos**: [Gr] *column*	Styloid process
		Styloglossus m
		Stylomastoid foramen

Notes, Links and Non-Anatomical Usages

Sponge, spongy. The corpus spongiosum is unpaired porous erectile tissue containing the urethra in males. It ends as the glans penis and lies ventrally along the penis

Stapes (the ear ossicle is a perfect little stirrup!), stapedial

The stapedius m damps down excessive movements of the stapes caused by loud sounds. If this muscle, or its nerve supply, are affected then acoustic disturbances may follow. Damage to which cranial n may result in hyperacusia?

Stellar, constellation (all stars together – just like Oscars night!). The stellate ganglion is near the neck of the first rib and is formed by the fusion of the inferior cervical and first thoracic ganglia of the sympathetic trunk

Stratified (layered). A stratum usually refers to some sort of layer such as the deep germinative layer of the epidermis

Lateral and medial olfactory striae arise from each olfactory tract just in front of the anterior perforated substance of the brain and optic chiasma

A frontal section through the junction of the calcarine and parieto-occipital sulci of the occipital cortex reveals a stripe of white matter embedded in the grey matter. The stripe reflects the organisation of the visual cortex and this area is called the striate cortex

So-called because the striped appearance (due to the regular organisation of actin and myosin filaments) is particularly noticeable histologically in this type of muscle

Grey matter extends in strips between putamen and caudate nucleus and appears striated in a frontal slice

The term stroma is often used to refer to the connective tissue component of an organ or part thereof

The styloid process is shaped like a little column

The muscle attaches to the styloid process and is the smallest and shortest of three muscles which do so. What are their nerve supplies?

A small foramen between the styloid and mastoid processes of the temporal bone. Part of the facial n leaves this foramen and radiates to supply the various muscles of facial expression

Anatomical Names	Latin/Greek or Other Origins	Examples
Sulcus, sulci	**Sulcus**: [L] *furrow, rut, little ditch, groove*	Cingulate sulcus
		Sulcus terminalis (tongue, R atrium)
Superficial[is]	**Superficies**: [L] *surface*; **superficialis**: [L] *related to the surface*	Flexor digitorum superficialis m
Supination, supinator	**Supinus**: [L] *lying on the back*	Supination
		Supinator m
Sural	**Sura**: [L] *calf of the leg*	Sural n
Suture	**Sutura**: [L] *seam, join*	Cranial sutures
Symphysis	**Syn-**: [Gr] *together*; **physis**: [Gr] *growth*	Pubic symphysis
Synchondrosis	**Syn-**: [Gr] *together*; **chondros**: [Gr] *cartilage*	Synchondrosis
Syndesmosis	**Syn-**: [Gr] *together*; **desmos**: [Gr] *bond, chain*	Syndesmosis

Sulcate (grooved). The cingulate sulcus partially girdles the corpus callosum on the medial surface of the cerebrum

Both grooves act as boundaries. In the tongue, it is between the embryologically distinct anterior and posterior parts; in the right atrium, it is also between different embryological parts

Remember that, in Anatomy, superficial always signifies closer to the skin surface. A superficial person is "only skin deep"…or less! Flexor digitorum superficialis m is a superficial finger flexor. It is supplied by the median n. Does it flex anything else apart from the finger joints?

Supine, supinate. In supination, the forearm bones lie roughly parallel to each other

One of the supinators of the forearm. It is supplied by the radial n and acts best when the elbow is extended. What acts best as a supinator in flexion at the elbow?

The sural n (spinal levels L5-S2) arises from the tibial division of the sciatic n but is joined by a communicating branch of the common peroneal n. What is its distribution?

These are non-synovial fibrous joints between various cranial bones. The skull of a newborn displays incompletely-fused sutures called anterior and posterior fontanelles. The former closes by 9-18 months and the latter by by 1-2 months. Why might you be interested in signs of bulging or depression of fontanelles?

When the two pubic bones meet and fuse together, they make the pubic symphysis. Symphyseal or symphysial

A cartilaginous joint. A continuous joint in which the bones are joined together by fibrocartilage or hyaline cartilage. Being non-synovial, there is no joint cavity and no synovial fluid. Where might you find examples of this type of joint?

A continuous, non-synovial type of joint in which the bones are joined by fibrous tissue (e.g. skull sutures and the interosseous membrane joining the tibia and fibula or the radius and ulna). From the second element, we also get desmosome (by which adjacent cells are bonded together)

Anatomical Names	Latin/Greek or Other Origins	Examples
Synostosis	**Syn-**: [Gr] *together*, **osteon**: [Gr] *bone*	Synostosis
Synergist, synergism	**Syn-**: [Gr] *together*, **ergon**: [Gr] *work*	Synergistic muscle
T		
Taenia, taeniae	**Taenia**: [L] *band, ribbon, narrow strip*	Taenia[e] coli
Talus, talar, tali, talo-	**Talus**: [L] *ankle*	Talus
		Tibiotalar lig
		Sustentaculum tali
		Talocalcaneona-vicular joint
Tarsus, tarsal, tarso-	**Tarsos**: [Gr] *flat of the foot, instep, edge of eyelid*	Tarsal bones
		Tarsometatarsal joints
		Tarsal plates or tarsi
Tectorial	**Tectorium**: [L] *a covering*	Tectorial membrane

Notes, Links and Non-Anatomical Usages

Another continuous, non-synovial joint. No prizes for guessing what joins the adjacent bones together! Examples?

Synergy, synergistic, energy (that which does work). Synergistic muscles act together to produce a given movement

Taeniasis (tapeworm infestation), taeniafuge (a drug for getting rid of tapeworms). Taeniae coli are three thin strips of smooth muscle running along the colon long axis

The tarsal bone with which the tibia and fibula articulate to produce the ankle joint. The talus is also the key of the medial longitudinal arch of the foot

The strong medial (deltoid) collateral lig of the ankle joint has tibionavicular, tibiocalcaneal and tibiotalar components. The deltoid lig is so strong that the medial malleolus of the tibia may break before the ligament

Literally, the support of the talus, a shelf of bone projecting from the medial side of the calcaneus

This is a multi-axial compound joint allowing gliding and rotational movements. Inversion and eversion of the foot involve this joint

Tarsus, metatarsus, metatarsal. The seven bones of the tarsus (what are their names?) form the ankle and bony arches of the foot. Tarsiers are insectivorous Primates with very long tarsal bones

These are roughly plane synovial joints between the metatarsal bones and the distal (cuneiform and cuboid) tarsal bones

Flat plates of connective tissue which strengthen the upper and lower eyelids

The tectorial membrane of the vertebral column is a continuation of the posterior longitudinal lig and inserts on the occipital bone. It forms a roof over the dens of the atlas. Another tectorial membrane is made of gelatinous material and forms a covering over the hair cells of the organ of Corti of the inner ear

Anatomical Names	Latin/Greek or Other Origins	Examples
Tectum	**Tectum**: [L] *roof, ceiling, cover*	Tectum
Tegmen, tegmentum	**Tegmen[tum]**: [L] *covering, shelter, roof*	Tegmentum
		Tegmen tympani
Tela	**Tela**: [L] *web, warp, loom*	Tela choroidea
Temporal, temporalis, temporo-	**Tempora**: [L] *the temples of the head*	Temporal bones
		Temporal lobe
		Temporal or temporalis m
		Temporomandibu- lar joint
Tension, tensor	**Tendere**: [L] *to stretch, spread, strain*	Tensor (fasciae latae, tympani, veli palatini) mm
Tentorium	**Tentorium**: [L] *tent*	Tentorium cerebelli
Teres	**Teres**: [L] *long and round, cylindrical*	Teres minor m

Detective (someone who uncovers things!). The tectum or roof of the midbrain bears the corpora quadrigemina

Integument, tegument. The tegmen[tum] of the midbrain is separated from the crus cerebri by substantia nigra. Together, these three components make the cerebral peduncle

The tegmen tympani is a thin plate of temporal bone which forms a roof for the mastoid antrum, tympanic cavity and canal of the tensor tympani m

Tela refers to web-like structures. The tela choroidea of the 3rd and 4th ventricles comprises pia mater with a highly vascular choroid plexus. What is their arterial supply?

Temporal bones underlie the temporal region

The temporal lobes lie in the middle cranial fossa. With what special senses are they associated and what is their arterial supply?

The muscle arises from the bone of the temporal fossa and its fibres converge to descend deep to the zygomatic arch to insert on the coronoid process of the mandible. The muscle is supplied by the mandibular division of cranial V n. What are its actions?

This is a bicondylar ellipsoid joint between the temporal articular tubercle and anterior part of the mandibular fossa (above) and mandibular condyle (below). An articular disc divides the joint into upper and lower parts. The complex allows elevation, depression, protrusion, retraction and side-to-side (chewing) movements. What muscles produce these actions?

Tension, tensile, tensor (all these muscles stretch things, i.e. the fascia lata of the thigh, tympanic membrane and soft palate respectively). A tendon takes the strain!

Tent. The tentorium cerebelli is a sheet of dura mater overlying the cerebellum in the posterior cranial fossa

Teres muscles tend to be rounded or cylinder-like. Teres minor and major mm produce movements at the shoulder joint. Which ones?

Anatomical Names	Latin/Greek or Other Origins	Examples
		Pronator teres m
		Ligamentum teres
Tertius	**Tertius**: [L] *third*	Peroneus tertius m
Testis, testicle, testicular	**Testis**: [L] *witness, testicle*	Testis
		Testicular artery
Thalamus, thalamic	**Thalamus**: [L] *inner room, bedroom*	Thalamus
		Interthalamic adhesion
Thenar, hypothenar	**Thena**: [Gr] *palm of the hand*	[Hypo]thenar mm [Hypo]thenar eminence
Thyroid, thyro-	**Thyreos**: [Gr] *oblong shield*; **eidos**: [Gr] *shape*	Thyroid cartilage
		Thyrocervical trunk
		Thyroidea ima a

Notes, Links and Non-Anatomical Usages

The pronator teres is a relatively long cylinder arising by humeral and ulnar heads between which runs the median n. The muscle attaches to the midshaft of the radius on its lateral side. Are there any other pronators to assist it?

The free edge of the falciform lig contains the round lig which, before birth, was the left umbilical v. After birth, it becomes a fibrous cord called the ligamentum teres or round lig of the liver. Other round ligaments are associated with the ovary and uterus. What do they represent?

Tertiary. There are three peroneal mm in each set and this is the third in importance. Peroneus tertius m is anatomically more related to the extensor digitorum longus m and is a dorsiflexor of the foot

Testify, testimony, intestate. The presence of testes used to be taken as proof of virility! Lack of them disqualified a man (such as a eunuch) from being called as a witness in a court case

The arteries arise from the anterior aspect of the abdominal aorta. What are the courses and relations of the left and right testicular aa?

The thalamus is a major inner region of grey matter in the cerebrum and a relay station but not for olfaction

Joins the left and right thalamus

The thenar eminence actually forms the fleshy part of the palm of the hand at the base or ball of the thumb. The thenar mm are found within this eminence. The hypothenar eminence is the less prominent (hypo-indicates below) medial or ulnar fleshy ridge running from wrist to little finger. The hypothenar mm underlie this eminence

Thyroxine. The thyroid cartilages 'shield' the larynx anteriorly

This arterial trunk arises from which part (1st, 2nd or 3rd) of the subclavian a? Its branches (inferior thyroid, transverse cervical and suprascapular aa) supply the inferior part of the thyroid gland, structures in the neck and muscles on the dorsal surface of the scapula

Ima: [L] *lowest*. Hence, the lowest of the arteries supplying the thyroid gland. The others are?

Anatomical Names	Latin/Greek or Other Origins	Examples
Tibia, tibiae, tibial[is], tibio-	**Tibia**: [L] *shinbone, flute*	Tibia
		Tibial n
		Tibialis posterior m
		Tibiofibular joints
Trabecula, trabeculae, trabecular	**Trabecula**: [L] *little beam, strut*	Trabeculae carneae
		Trabecular bone
Trachea, tracheal, trachealis	**Trachos**: [Gr] *rough*	Trachea
		Tracheal cartilages
		Trachealis m

Notes, Links and Non-Anatomical Usages

The medial long bone of the leg forming the shin

This is one of the two main divisions of the sciatic n. It pursues a straight course on the posterior aspect of the leg, first on fascia covering the tibialis posterior m and then on the tibia itself. At the distal end of the tibia, it runs deep to the flexor retinaculum and behind the medial malleolus to divide into the medial and lateral plantar nn of the foot

The deepest of the deep muscles of the posterior compartment of the leg. It arises from the interosseous membrane and adjacent bones and passes behind the medial malleolus of the tibia to insert where in the foot? Remember, it is the main invertor, and a powerful plantar flexor, of the foot

The proximal joint (between the lateral condyle of the tibia and head of the fibula) is almost plane. The distal joint is a syndesmosis. What ligaments are associated with these bones and joints?

Trabeculae carneae are muscular roughenings of the ventricular walls of the heart covered in endocardium. They are better developed in the left ventricle. They appear as irregular ridges, columns, bands or protrusions

In contrast to compact bone, sections of trabecular bone show spaces criss-crossed by bars or struts whose patterns are related to the lines of stress to which the bones are exposed

Actually an abbreviation of trachea arteria (rough artery) so somebody regarded it as an artery but had noticed that it was rougher than normal (due, clearly, to the many C-shaped rings of cartilage!). At what level does the trachea bifurcate into the main bronchi? Trachoma

The trachea is kept patent by C-shaped rings of cartilage which are deficient posteriorly. The lowest cartilage has an inferior ridge (the carina) which runs downwards and backwards between the left and right main bronchi. The isthmus of the thyroid gland joins the two lobes and crosses the second and third tracheal cartilages

A transverse band of smooth muscle which runs between the posterior ends of tracheal cartilages and lies between the trachea and oesophagus

Anatomical Names	Latin/Greek or Other Origins	Examples
Tract, tractus	**Tractus**: [L] *line/ stroke, drawing, region, course*	Tractus solitarius
		Alimentary tract
		Iliotibial tract
Trapezius, trapezium, trapezoid	**Trapezion**: [Gr] *table*; **eidos**: [Gr] *shape*	Trapezius m
		Trapezium
		Trapezoid
Triceps	**Triceps**: [L] *three-headed*	Triceps brachii m
Tricuspid	**Tricuspid**: [L] *having three points*	Tricuspid valve
Trigeminal	**Trigeminus**: [L] *threefold, triple, triplet*	Trigeminal n
Trigone	**Trigonus**: [L] *triangular*	Trigone
Triquetral	**Triquetrus**: [L] *three-cornered, triangular*	Triquetral bone

Notes, Links and Non-Anatomical Usages

Tract, tractor. Often used to refer to a bundle of nerve fibres or a pathway in the spinal cord or brain. For instance, the tractus solitarius (solitary or lonely tract) lies in the medulla oblongata and comprises primary sensory fibres and descending fibres from cranial nn VII, IX and X. These include gustatory (taste) fibres

Tract is also used to refer to the course of a major passage or system. So the alimentary (digestive) tract runs from the oral cavity to the anus

A longitudinal fibrous strengthening of the fascia lata of the thigh. With its associated muscles (what are they?), it is a flexor, abductor and medial rotator at the hip joint and contributes to knee stability. What are its attachments?

A trapezium is a quadrilateral with one pair of sides or with no sides parallel. The latter more closely describes the shape of the muscle. Trapeze (usually quadrilateral)

A carpal bone distinguishable by its shape. It has a tubercle to which the flexor retinaculum attaches (so the retinaculum goes from the hook of the hamate to the tubercle of the trapezium!)

A bone between the capitate and trapezium bones of the wrist

The three heads of the triceps m are long, medial and lateral and their fibres run together to form a tendon which inserts on the olecranon. What are the origins of the three heads?

The valve has three cusps which form the atrioventricular valve on the right side of the heart. How do the papillary mm and chordae tendineae help this valve to close during systole?

Cranial V n. The triplet of trigeminal branches: the ophthalmic, maxillary and mandibular divisions

Trigonal, trigonometry. The trigone of the bladder is a smooth triangular area with the internal urethral and two ureteric orifices at its angles

A triangular carpal bone on the distal palmar surface of which sits the pisiform bone

149

Anatomical Names	Latin/Greek or Other Origins	Examples
Trochanter, trochanteric	**Trekhein**: [Gr] *to run, roll;* **trokhos**: [Gr] *wheel*	Greater and lesser trochanters
		Intertrochanteric crest
Trochlea, trochlear	**Trochlea**: [L] *pulley*	Trochlea of the humerus
		Trochlear n (cranial IV)
Tuber, tubercle, tuberosity	**Tuber**: [L] *lump, swelling, protuberance*	Omental tuber
		Infraglenoid tubercle
		Deltoid tuberosity
Tunica, tunicae	**Tunic**: [L] *coat*	Tunica vaginalis testis
Turbinate	**Turbineus**: [L] *like a spinning top*	Turbinates
Tympanic, tympano-	**Tympanum**: [L] *drum, tambourine*	Tympanic membrane
		Epitympanic recess

Notes, Links and Non-Anatomical Usages

The term trochanter was originally applied to the rounded head of the femur which rotates, like a wheel, in the acetabulum. Nowadays, it refers to the two bony processes of the proximal end of the femur. What muscles and ligaments attach to the greater and lesser trochanters?

The bony ridge on the posterior aspect of the femur which runs between the two trochanters. The quadrate tubercle lies on this crest

The trochlea of the distal end of the humerus is shaped like a pulley and articulates with a notch on the ulna

The trochlear n supplies the superior oblique m of the eyeball. This muscle has its line of pull altered by a small fibrous loop, the trochlea, on the frontal bone at the superomedial angle of the orbit

Tuber, protuberance, tubercular, tuberosity. The omental tuber is on the inferior or visceral surface of the liver to the right of the gastric impression and to the left of the caudate lobe. The slight prominence is so-called because it is in contact with the lesser omentum

A lump related to the inferior part of the rim of the glenoid fossa of the scapula. To it is attached the long head of triceps brachii m. It is worth noting that the long head of biceps brachii m runs to a comparable supraglenoid tubercle!

A roughened area of bone on the middle of the lateral aspect of the humeral shaft into which the tendon of the deltoid m inserts

Tunicate (marine animal with a body surrounded by a jelly-like coat). The tunica vaginalis of the testis is a serous sac invaginated by the testis and lying deep to the three layers of spermatic fascia. Its visceral and parietal layers are continuous posteriorly

Turbine. Turbinate has the sense of curving or scroll-like or shell-like and refers to the curved shelves of bone projecting from the lateral walls of the nasal cavities

Tympani (set of kettledrums). The tympanic membrane is often called the eardrum

The epitympanic (above the eardrum) recess is the part of the tympanic cavity lying above the tympanic membrane

Anatomical Names	Latin/Greek or Other Origins	Examples
		Tympanosqua-mous fissure
U		
Ulna, ulnar, ulnaris	**Ulna**: [L] *elbow, arm*	Ulna
		Ulnar n
		Flexor carpi ulnaris m
Umbilicus, umbilical, umbo	**Umbo**: [L] *boss of a shield;* **umbilicus**: [L] *little boss or navel*	Umbilicus
		Paraumbilical vv
		Umbo
Uncus, uncinate	**Uncus**: [L] *hook, crooked, curved*	Uncus
		Uncinate process

Notes, Links and Non-Anatomical Usages

Sometimes called the squamotympanic fissure, this narrow fissure separates the superior tympanic part of the temporal bone from the posterior part of the articular portion of the mandibular fossa

Actually the medial long bone of the forearm and not of the arm

Arises from the medial cord of the brachial plexus and supplies muscles in the forearm and hand and skin on the medial (ulnar) side of the hand and medial two fingers

One of only two muscles in the forearm which are innervated by the ulnar n. This muscle is a flexor of the wrist and lies on the medial (ulnar) side of the wrist. The pisiform bone develops as a sesamoid bone in its tendinous insertion

Umbilicus (the little boss of the belly, 'belly-button'), umbilical. What spinal level is represented by the dermatome which includes the umbilicus? An alternative way of dividing the abdomen is to describe four quadrants via a vertical and a horizontal line, each of which passes through the umbilicus. The plane corresponding to the horizontal line (the transumbilical plane) passes through the disc between vertebrae L3 and L4

Passing 'alongside the umbilicus', these veins connect veins of the anterior abdominal wall with those of the left portal v extending along the ligamentum teres in the falciform lig. It is these paraumbilical vv which distend to form the caput Medusae in portal hypertension

The rounded end of the handle of the malleus where it attaches to the tympanic membrane. When examined with an otoscope (= auriscope), a light reflection – 'the cone of light' – runs anteriorly and inferiorly from the umbo

The uncus is a hook-like part of the hippocampal gyrus

The hook-shaped part of the pancreatic head that lies posterior to the origin of the superior mesenteric vessels and arises from the ventral pancreas

Anatomical Names	Latin/Greek or Other Origins	Examples
Ureter, ureteric	**Ourein**: [Gr] *to urinate*	Ureter
		Ureteric constrictions
Urethra, urethral	**Ourein**: [Gr] *to urinate*	Urethra
		Urethral sphincter m
Uro-	**Oura**: [Gr] *urine*; **ourein**: [Gr] *to urinate*	Urogenital triangle
		Urothelium (urinary or transitional epithelium)
Utricle, utriculus	**Utriculus**: [L] *little bag*	Prostatic utricle
		Utriculus of the inner ear
Uvula, uvulae, uvular	**Uva**: [L] *grape*; **uvula**: [L] *little grape*	Uvula (palate, cerebellum)

V

Anatomical Names	Latin/Greek or Other Origins	Examples
Vagina, vaginal[is]	**Vagina**: [L] *sheath, scabbard*	Vagina
		Vaginal a

Notes, Links and Non-Anatomical Usages

The ureters pass urine from kidney to bladder

Constrictions found typically at three sites. Where are they?

The urethra passes urine from bladder to outside the body

This sphincter is voluntary (after early infancy!) and controls the passage of urine

Urine, urinal, urinary, urinate, etc. Generally, uro- implies related to the urinary system. The urogenital triangle is the anterior region of the perineum, the posterior being the anal triangle. The anterior triangle is defined by lines joining the pubic symphysis and two ischial tuberosities

Transitional epithelium is stratified but the apparent number of layers varies according to the degree of stretch or distension

The prostatic utricle opens on an elevation called the colliculus seminalis on the posterior wall of the male urethra. Embryologically, the utricle corresponds to the uterus and vagina

The utricle or utriculus of the inner ear is the larger of two sacs (the other being the saccule or sacculus) lying in the vestibule of the membranous labyrinth

The uvula of the soft palate is a median conical process which hangs inferiorly from the posterior border. It may be elevated and retracted by a bilateral muscle (the musculus uvulae) which is supplied by the pharyngeal plexus. The uvula of the cerebellum is part of the inferior cerebellar surface between the pyramid and nodule

A sheath or scabbard accommodates the blade of a knife, sword, etc. No prizes for guessing what the anatomical 'blade' might be!

This is a branch of the internal iliac a. It descends in front of the ureter to the base of the broad lig and, at the lateral vaginal fornix, crosses superior to the ureter. It supplies branches to the vagina, cervix, uterus and uterine tubes

Anatomical Names	Latin/Greek or Other Origins	Examples
		Pubovaginalis m
Vagus, vagi, vagal	**Vagus**: [L] *wandering, rambling*	Vagus n (cranial X)
		Vagal nuclei
		Vagal trunks
Vallecula, valleculae	**Vallis**: [L] *valley*	Vallecula epiglottica
Varicose, varicosity	**Varicosus**: [L] *full of swollen veins*	Varicose vv
Vas, vasa, vaso-	**Vas**: [L] *vessel, vase*	Ductus (vas) deferens
		Vasa vasorum
		Vasodilation
Vastus	**Vastus**: [L] *monstrous, great, desolate, waste*	Vastus medialis m

Notes, Links and Non-Anatomical Usages

This muscle constitutes part of the levator ani m in females and acts as a vaginal sphincter muscle. The pubic part of levator ani is divisible into three sets of fibres that run downwards, posteriorly and medially. The medial part is pubovaginalis m. What are the other parts called?

Vagabond (no fixed abode), vagary (erratic idea or action), vagrant (a wanderer), vague (not fixed, absent-minded). The vagus n is so-called because of its extensive wandering course innervating structures between the cranium and midgut

The vagus n has four nuclei in the medulla oblongata: the dorsal nucleus (visceral efferent, parasympathetic), nucleus ambiguus (somatic efferent), nucleus solitarius (visceral afferent) and spinal trigeminal nucleus (somatic afferent)

The anterior vagal trunk is formed mainly from the left vagus n and enters the abdomen on the anterior surface of the oesophagus and sends branches to the stomach, pylorus and liver. The posterior vagal trunk is formed mainly from the right vagus n on the posterior of the oesophagus. Branches supply the stomach, coeliac and superior mesenteric plexuses, and intestines as far as the splenic (left colic) flexure

Vallecula (literally, little valley). The vallecula epiglottica on each side forms a valley between the median and lateral glossoepiglottic folds. These mucosal folds run between the tongue and epiglottis

So…it's the legs, not the veins, that are varicose!

Vase, vasomotor, vasopressin. The name vas deferens has been superseded by ductus deferens so that the vas/vaso- root can be confined to blood and lymphatic vessels. The ductus deferens runs between the epididymis and common ejaculatory duct via the superficial inguinal ring, inguinal canal and deep inguinal ring

Vasa vasorum (vessels of the vessels)

Dilation of a blood vessel. This and vasoconstriction are brought about by vasomotor fibres innervating smooth muscle in the vessel wall

Vast, devastation. The vastus medialis m arises from the intertrochanteric line and linea aspera of the femur. It inserts into the quadriceps tendon and patella. It is supplied by the femoral n. What are its actions?

Anatomical Names	Latin/Greek or Other Origins	Examples
Vein, vena, venae, venous	**Vena**: [L] *vein, blood vessel, way*	Internal jugular v
		Portal venous system
		Venae cavae
Velum, veli	**Velum**: [L] *sail, curtain*	Levator veli palati[ni] m
		Tensor veli palate[ni] m
Ventral	**Venter**: [L] *belly, stomach, protuberance*	Ventral
Ventricle, ventricular	**Ventriculus**: [L] *small belly, stomach*	Ventricle
		Atrioventricular node
Vermis, vermiform	**Vermis**: [L] *worm*	Vermis of the cerebellum
		Vermiform appendix
Vertebra, vertebrae, vertebral	**Vertebra**: [L] *a joint, especially of the backbone*	Vertebra

Notes, Links and Non-Anatomical Usages

Vein, venous, venesection. The internal jugular v begins at the jugular foramen as a continuation of the sigmoid sinus. It descends in the carotid sheath (which contains what other important structures?) and ends by joining the subclavian v to form a brachiocephalic v

The tributaries of the portal v are the splenic, inferior mesenteric, superior mesenteric, gastric (left and right) and cystic vv

Both the SVC and IVC enter the right atrium of the heart. What areas do they drain?

The word veil comes from this root. The muscle is tested when you ask a patient to say 'Ah'. If both are working normally, the soft palate is raised. If the muscle on one side is non-functional, the soft palate deviates towards the other side

Unlike the above muscle, this one passes round the pterygoid hamulus and stretches the soft palate

Ventriloquist. Ventral and anterior are often used interchangeably

The cerebral and cardiac ventricles are obvious examples but they are not so-named because they are stomach-shaped

The node is found in the lower part of the interatrial septum just above the attachment of the septal cusp of the right atrioventricular valve of the heart

Vermicelli (looks, but doesn't taste, like worms!), vermiform, vermifuge (a drug which kills intestinal worms), Vermouth comes from the shrub wormwood (absinth, a potent green alcoholic drink had a high content of wormwood but was banned because of toxic effects. It is now becoming fashionable again!). The vermis of the cerebellum runs longitudinally between the two hemispheres

The vermiform appendix is aptly named

Ultimately, from **vertere**: [L] *to turn*. The vertebral joints also allow flexion and extension movements. Aversion, version, vertigo

Anatomical Names	Latin/Greek or Other Origins	Examples
		Vertebral aa
Vertex	**Vertex**: [L] *crown of the head, top, summit*	Vertex
Vesicle, vesicular, vesical, vesico-	**Vesica**: [L] *bladder, ball*	Seminal vesicle
		Vesical plexus
		Vesico-uterine pouch
Vestibule, vestibular, vestibulo-	**Vestibulum**: [L] *porch, entry, vestibule*	Vestibule (inner ear, larynx, nose, vulva)
		Vestibulocochlear n (cranial VIII)
Vinculum, vincula	**Vinculum**: [L] *bond, tie*; **vincula**: [L] *chains*	Vinculum
Viscus, viscera, visceral	**Viscus**: [L] *internal organ, intestines, bowels, inmost part*	Viscus
		Visceral pericardium
Vitreous	**Vitreus**: [L] *glassy*	Vitreous body
Vomer	**Vomer**: [L] *ploughshare*	Vomer

Notes, Links and Non-Anatomical Usages

Arise from which part of the subclavian aa? Each vertebral a runs through the foramina transversaria of cervical vertebrae (except C7) and enters the skull via the foramen magnum. What artery is formed when the two vertebral aa unite?

Anatomically, this is the most superior point of the cranium. The Latin word also had the meaning of 'whirlpool or vortex' from which we get the word vertigo. The term vertical is from the same origin and refers to a line or plane in which the top lies directly above the bottom

Each seminal vesicle develops from the ampulla of the ductus deferens with which it forms the ejaculatory duct. The vesicles can be palpated anteriorly via the anal canal

Pelvic autonomic nn distributed to the bladder

In females, peritoneum is reflected from the anterior of the uterus to the superior surface of the bladder to create a shallow pouch

The vestibules of the inner ear, larynx and nasal cavities are merely entrances to those spaces. In females, the vestibule is also the space between the labia minora containing the external openings of the vagina, urethra and lesser vestibular glands

Leaves the middle cranial fossa via the internal auditory (acoustic) meatus to supply organs of hearing (audition) and balance (equilibration)

Vincula are found at various sites but all are 'connective' tissues such as those which attach flexor digitorum tendons to the phalanges

Eviscerate (disembowel). Visceral and splanchnic both refer to the organs of the body

Serous membranes usually have visceral (covering the organ) and parietal (lining the walls) components separated by a fluid-filled sac. Visceral pericardium covers the heart and corresponds to epicardium

Vitrify (to turn to glass). Vitreous humour is found in the vitreous body which accounts for about 80% of eyeball volume. It forms a colourless, transparent gel of somewhat glassy appearance

The vomer is a ploughshare-like bone forming the postero-inferior part of the nasal septum

Anatomical Names	Latin/Greek or Other Origins	Examples
Vulva	**Vulva**: [L] *wrapper, womb*	Vulva

X

Xiphisternum or xiphoid process	**Xiphos**: [Gr] *sword*; **eidos**: [Gr] *shape*	Xiphisternum Xiphoid process

Z

Zygoma, zygomatic	**Zygos**: [Gr] *yoke*	Zygoma, zygomatic bone

Notes, Links and Non-Anatomical Usages

Vulvitis. Vulva is a collective term, like pudenda, used to describe the mons pubis, labia, clitoris, vestibule and vestibular glands of the female genitalia. So it actually excludes the womb!

The two names are interchangeable. The manubrium (handle), body (blade) and xiphoid process (tip) together resemble a short sword favoured by Roman and Greek soldiers

Names used to identify the bones which give the prominences to the cheeks and form the inferolateral margins of the orbits. The reason for their likeness to a yoke is possibly related to the fact that they join several other bones of the skull (which ones?). In the same sense, a zygote comprises a yoked or paired set of gametes

Appendix
Terms organised by body region

Head & Neck:

Abduction, abducent, abductor 14
Accessory 14
Acinus, acini, acinar 14
Acoustic 16
Adenoid, adenoidal, adeno- 16
Aditus 18
Ala, alae, alar 18
Alveolus, alveoli, alveolar 18
Ampulla[e] 18
Amygdala, amygdaloid 18
An[n]ulus, annular 20
Ansa 20
Antrum, antra, antral 20
Aqueous 22
Arachnoid 22
Arytenoid, ary- 24
Atlas, atlanto- 24
Audition, auditory 26
Auricle, auricular[is], auriculo- 26
Axis, axial 28

Base, basal, basilar, basilica 28
Bicuspid 28
Bregma 30
Bucca, buccae, buccal, bucco- 30
Buccinator 30
Bulb, bulbo- 30
Bulla[e] 32

Caecum, caecal 32
Calvaria 34
Canine 34
Canthus, canthi 34
Caput, capitis, capitate 34
Carotid 36
Caruncle, caruncular 38
Cauda, caudal, caudate 38
Cavernous 40

Cephalon, cephalic 40
Cerebellum, cerebellar 40
Cerebrum, cerebral, cerebro- 40
Cervix, cervical[is] 42
Chiasma, chiasmata 42
Choana[e] 42
Chondral, chondro- 42
Chorda[e] 42
Choroid 42
Ciliary 42
Cingulate 44
Circumvallate 44
Cisterna, cisternae, cisternal 44
Claustrum 44
Clinoid 46
Clivus 46
Cochlea, cochleae, cochlear 46
Coeruleus 46
Colliculus, colliculi 48
Commissure, commissural 48
Concha[e] 50
Condyle, condylar 50
Conjunctiva, conjunctival, conjoint 50
Cornea, corneal 52
Cornu, cornua 52
Coronary, coronoid, coronal 52
Corpus, corpora 52
Cortex, cortical, cortico- 52
Cranium, cranial, cranio- 54
Cribriform 54
Cricoid, crico- 54
Crista, cristae, cristal 54
Crus, crura, crural 54
Crux, cruciate, cruciform 56
Cuneus, cuneate, cuneiform 56
Cupola 58

Decidua, deciduous 58
Decussation 58

Upper Limb:

Thorax:

Pelvis/Perineum:

Miscellaneous:

Index

Index

Index